Advances and Challenges in Microplastics

Edited by El-Sayed Salama

Published in London, United Kingdom

Advances and Challenges in Microplastics
http://dx.doi.org/10.5772/intechopen.104368
Edited by El-Sayed Salama

Contributors
Tore Syversen, Grethe Lilleng, Raffaella Mossotti, Giulia Dalla Fontana, Anastasia Anceschi, Tiziano Battistini, Enrico Gasparin, Mikail Olam, Paul Agbekpornu, Isaac Kevudo, Ahmad K. K Jassim, Terwase Wuave, Ahmed Sabo, Dalia Saad, Nelly Sapojnikova, Aisha S. Alwuhaib, Vitaly Zinkevich, Tamar Kartvelishvili, Nino Asatiani, Carolina Peña-Montes, Karla Lizzeth Morales-Cano, Leandro Hermida-Castellanos, Luis Alberto Peralta Peláez, Carlos M. Adame-Adame, Assad A Ahmed Al-Thukair, Hamza Jawad Al-Shaikh Ali, Abdurahiman Kambrath Pulikkoden, Basheer Chanbasha

First published in London, United Kingdom, 2023 by IntechOpen
IntechOpen is the global imprint of INTECHOPEN LIMITED, registered in England and Wales, registration number: 11086078, 5 Princes Gate Court, London, SW7 2QJ, United Kingdom
Printed in Croatia

British Library Cataloguing-in-Publication Data
A catalogue record for this book is available from the British Library

Additional hard and PDF copies can be obtained from orders@intechopen.com

Advances and Challenges in Microplastics
Edited by El-Sayed Salama
p. cm.

This title is part of the Environmental Sciences Book Series, Volume 8
Topic: Pollution
Series Editor: J. Kevin Summers
Topic Editors: Ismail Md. Mofizur Rahman and Zinnat Ara Begum

Print ISBN 978-1-83768-304-8
Online ISBN 978-1-83768-305-5
eBook (PDF) ISBN 978-1-83768-306-2
ISSN 2754-6713

For EU product safety concerns:
IN TECH d.o.o., Prolaz Marije Krucifikse Kozulić 3, 51000 Rijeka, Croatia,
info@intechopen.com or visit our website at intechopen.com.

IntechOpen Book Series

Environmental Sciences

Volume 8

Aims and Scope of the Series

Scientists have long researched to understand the environment and man's place in it. The search for this knowledge grows in importance as rapid increases in population and economic development intensify humans' stresses on ecosystems. Fortunately, rapid increases in multiple scientific areas are advancing our understanding of environmental sciences. Breakthroughs in computing, molecular biology, ecology, and sustainability science are enhancing our ability to utilize environmental sciences to address real-world problems.

The four topics of this book series - Pollution; Environmental Resilience and Management; Ecosystems and Biodiversity; and Water Science - will address important areas of advancement in the environmental sciences. They will represent an excellent initial grouping of published works on these critical topics.

Meet the Series Editor

J. Kevin Summers is a Senior Research Ecologist at the Environmental Protection Agency's (EPA) Gulf Ecosystem Measurement and Modeling Division. He is currently working with colleagues in the Sustainable and Healthy Communities Program to develop an index of community resilience to natural hazards, an index of human well-being that can be linked to changes in the ecosystem, social and economic services, and a community sustainability tool for communities with populations under 40,000. He leads research efforts for indicator and indices development. Dr. Summers is a systems ecologist and began his career at the EPA in 1989 and has worked in various programs and capacities. This includes leading the National Coastal Assessment in collaboration with the Office of Water which culminated in the award-winning National Coastal Condition Report series (four volumes between 2001 and 2012), and which integrates water quality, sediment quality, habitat, and biological data to assess the ecosystem condition of the United States estuaries. He was acting National Program Director for Ecology for the EPA between 2004 and 2006. He has authored approximately 150 peer-reviewed journal articles, book chapters, and reports and has received many awards for technical accomplishments from the EPA and from outside of the agency. Dr. Summers holds a BA in Zoology and Psychology, an MA in Ecology, and Ph.D. in Systems Ecology/Biology.

Meet the Volume Editor

El-Sayed Salama is a Professor of Environmental Health and the head of the Green Environmental and Energy Laboratory (GEEL), Lanzhou University (LZU), China. He was previously an assistant professor at Hanyang University (HYU), South Korea. He is also appointed as a Distinguished Expert in Science and Technology in the Ecological Industry of Gansu Province, China. He was a Brain Korea (BK21) fellow for his MSc and Ph.D. degrees and also performed Post-Doctoral at HYU. Prof. Salama is an editor, associate editor, and editorial board member of several international journals. His research team at LZU focuses on bioenvironmental sciences and bioenergy.

Contents

Preface

Worldwide, more than 400 million tons of plastic are produced annually. These plastics break down into small particles and those with a length of < 5 mm are called microplastics (MPs). The World Health Organization (WHO) has identified MPs as an emergency pollutant for the environment and human health. MPs hold toxic chemicals (such as plasticizers and colorings) that are used as ingredients during the polymerization process. Moreover, MPs act as transport vectors of other hazardous substances (pesticides, heavy metals, and biofouling), which can accumulate in the environment and induce toxicity to biodiversity. In addition to the inhalation of MPs, their existence in drinks and foods is the main pathway for entering the human body. The accumulation of MPs in the human body places people at risk for respiratory problems, cytotoxicity, and inflammatory as well as autoimmune illnesses. Therefore, it is crucial to identify the safest and most effective strategies for removing MPs from the environment. This book provides a comprehensive summary of MPs and discusses advances and challenges in removing them from our environment.

We would like to express our gratitude to all the contributing authors. We also wish to thank the publishing process manager at IntechOpen for being generously helpful.

El-Sayed Salama
Lanzhou University,
Lanzhou, China

Chapter 1

Why Microplastics Are Exceptional Contaminants?

Dalia Saad

Abstract

Due to the heterogeneous nature of the physiochemical properties of microplastics (MPs), their behaviour in the environment is quite complex compared to other contaminants. The variety of polymers, wide range of sizes, variable shapes and numerous colours influence their mobility, transport and distribution in the different environmental compartments. For example, different shapes and sizes are distributed differently, which influence their bioavailability and ecological impacts. The uptake of MPs by aquatic biota also depends, among others, on their characteristics. This book chapter aims to discuss the ecological and toxicological impacts of MPs in relation to their physical and chemical properties. The chapter starts with a brief introduction explaining the uniqueness of MPs as emerging contaminants and a driver of environmental change. The following two sections then provide deeper insights into their ecological impact at all levels of the ecosystem and highlight the complexity associated with their toxicological effects. Finally, the last section provides more discussion about their properties in the context of their environmental behaviour, fate, bioavailability and toxicity.

Keywords: microplastics, physiochemical properties, behaviour, bioavailability, toxicity

1. Introduction

Plastic pollution was one of the biggest environmental challenges until the discovery of microplastics (MPs) in the early 21st century. While plastics are easily visible and their environmental impacts are well documented, MPs are not visible and their ecological impacts are less understood [1].

MPs are exceptional pollutants with a broad range of individual properties. For instance, they are made of different polymers with different densities and chemical compositions (there are currently more than 5,300 types of synthetic polymers); they exist in variable shapes (fibres, fragments, foams, films, spheres, flakes, foils, sheets and granules) and are found in a wide range of sizes. These heterogeneous properties result in heterogeneous behaviour, fates and effects that are far more complex compared to other environmental pollutants. To add to this complexity, their properties and behaviour can also change over time, thus their ecological effects [1–3].

According to their physio-chemical properties, MPs are distributed differently in aquatic environments, which makes them available for uptake by a wide range of aquatic biota including plants. MPs are reported to interact with aquatic plants and accumulate into plants' tissues. This enables them to penetrate aquatic food webs at multiple trophic levels and ecological niches. Yet, the degree and type of effects that they cause when consumed by organisms depend on their properties including polymer type, size, shape and colour, as well as their constituent chemicals [4–8].

Due to their greater surface area, MPs have a propensity to adsorb other pollutants such as metals, pharmaceuticals and persistent organic pollutants (POPs). They also host pathogens, such as bacteria and viruses, thus, providing an additional pathway of exposure of aquatic species to contaminants. In other words, MPs can serve as a micro-vector for a mix of toxic chemicals and pathogens [9–14].

2. Ecological impact

Over the years, several studies across the globe have reported MPs in different environmental compartments including rivers, lakes, estuaries, oceans, harbours, groundwater and in the atmosphere, as well as in Antarctica. Once they enter the environment, their residence time lasts for decades due to their low degradation rates, resulting in long-lasting impacts [15–19].

In natural environments, MPs are exposed to a variety of degradation processes through different environmental conditions including weathering, biodegradation, oxidation, mechanical forces and phytodegradation. Phytodegradation of MPs is reported to produce greenhouse gases (GHGs), mostly, methane and ethylene, thus, contributing to climate change. The emission of GHGs by the degradation of MPs is relatively low, however, with continuous degradation, the same amount of MPs may release more GHGs over time [20–23].

In the atmosphere, MPs can be transported with winds around the earth. Airborne MPs may influence earth's climate by scattering and absorbing solar and terrestrial radiation, leading to atmospheric warming or cooling depending on particle size, shape and composition. However, the radiative effects of airborne MPs on climate are less understood [22, 24].

In marine environments, the widespread of MPs affects the light transmission, thereby influencing the efficiency of phytoplankton photosynthesis, which impacts both their growth and role in balancing the marine environment. Studies have shown that the photosynthetic rate of phytoplankton (*Dunaliella tertiolecta*) is reduced by 45% after being exposed to MPs. Additionally, MPs may influence the circulation of organic matter and nutrients, which affects the carbon stock of ocean [25].

In terrestrial ecosystems, MPs can cause significant environmental changes with potential consequences on soil function, plant growth, soil biota and microbial communities; ultimately, MPs have the potential to impact the biodiversity. When dispersing in the soil matrices, MPs form aggregates and cause alteration in the physical properties of the soil, including water holding capacity, soil bulk density and soil structures. For example, MPs can create channels for water movement in the soil, thus, accelerating the evaporation of soil water. This further leads to destruction in the soil structure, which may result in desiccation cracking on the soil surface. The impact of MPs on the soil is not limited to the physical properties, MPs can also affect soil chemistry, for instance, by altering the levels of dissolved organic carbon,

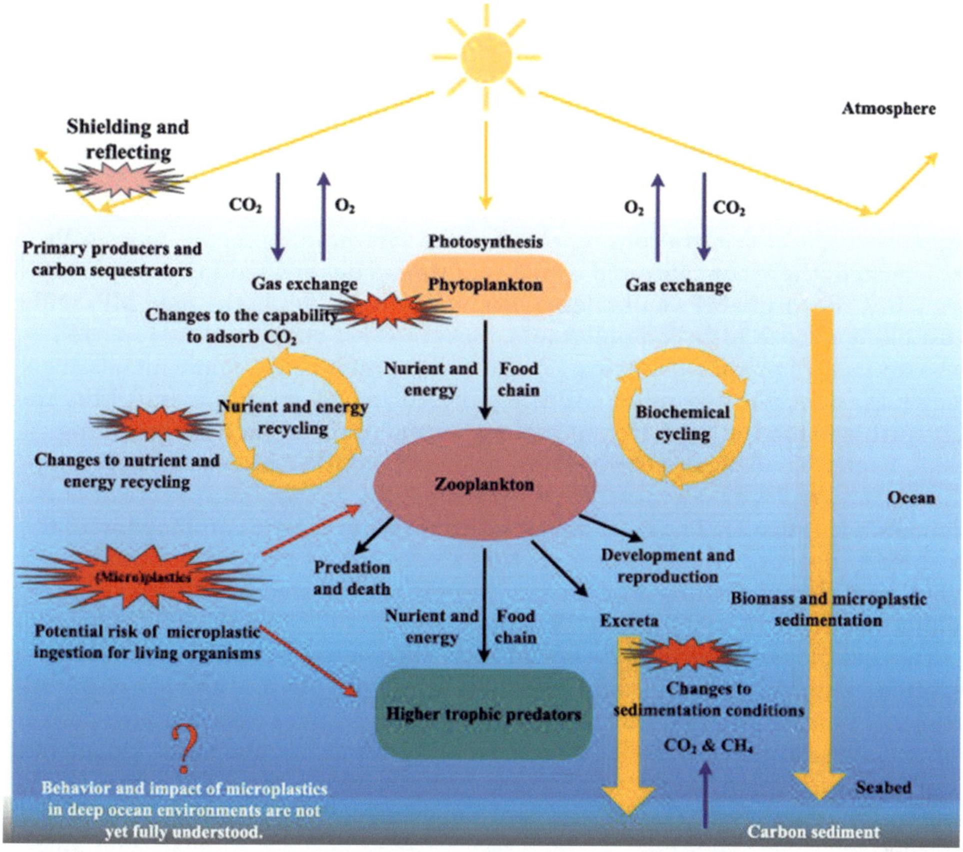

Figure 1.
Ecological impacts of MPs. Source: Shen et al. [20].

phosphorus, and nitrogen. This leads to changes in the nutrient cycling processes in the soil. There is also a growing body of evidence suggesting that MPs can affect soil-plant interactions, which in turn impacts plant growth. Several studies have reported significant changes in plant biomass, leaf and root traits and tissue elemental composition [26–36]. In short, MPs have profound effects on the ecosystem at all levels (**Figure 1**).

3. Toxicity

The toxicity of MPs comes from (i) their chemical constituents, which include both the polymers (polyaromatic hydrocarbon) and the chemical additives; (ii) the environmental pollutants adsorb onto their surfaces; (iii) pathogens colonized onto their surfaces.

During plastic processing and manufacturing, a variety of chemicals are added to enhance/adjust their properties and to make them into materials fit for intended purposes. Most of these chemicals are toxic and harmful to the environment, such as dyes, phthalates, flame retardants, pigments and stabilizers. Some of these additives tend not to be strongly bound within the matrix of the polymer and they can potentially desorb and be leached out into the host environment [37–39].

On the other hand, due to their small size and greater surface area, MPs have a tendency to adsorb wide range of contaminants from the surrounding media. Pollutants such as persistent organic pollutants (POPs), metals, pesticides and pharmaceuticals are readily bound to MPs. In natural environments, and depending on the prevailing environmental conditions, MPs may act as a sponge removing and/or concentrating these contaminants. It is reported that the concentrations of contaminants on the surface of MPs may reach up to 100-fold higher than the concentrations reported in the surrounding media. Once MPs are ingested, these concentrated contaminants can be released inside organisms. Arguably, the virgin MPs will release plastic additives, while the aged MPs will most likely release adsorbed pollutants. Most of these chemicals are reported to be toxic; for instance, POPs are known to be carcinogenic, while metals are known as endocrine disruptors. Additionally, in aquatic environments, MPs are susceptible to biofouling different pathogens/microbial organisms including fungi, bacteria and algae colonize MPs' surfaces and form biofilms. Therefore, MPs act as carriers or micro-vector for transporting a complex mixture of contaminants (**Figure 2**). The leaching of additives from plastic combined with the chemicals adsorbed to plastic renders MPs a 'cocktail' of toxic contaminants. When particles containing adsorbed chemicals are ingested by an organism, pollutants can be released [9–11, 38, 41–47].

The toxicological effects of the uptake of MPs by several aquatic biotas are reported in a variety of exposure studies, including both physical and bio-chemical changes. For instance, MPs were observed to cause oxidative stress, immune destruction and alterations in the level of enzyme activity, tissue morphologies, kidney functions, gene expression and the total protein and glucose. Further, MPs may inhibit weight gain and growth. This, in addition to physical changes, such as abnormally impaired movement coordination, increased respiration and abnormal swimming patterns [48–53].

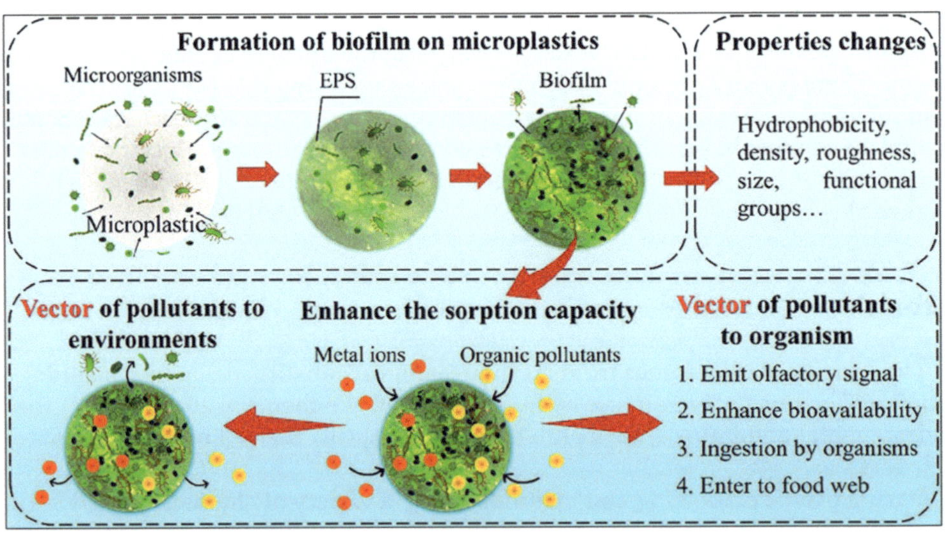

Figure 2.
Interaction of MPs with co-existing pollutants. Source: Wang et al. [40].

4. Characteristics of MPs: implications on their behaviour, bioavailability and toxicity

The unique nature of MPs is clearly illustrated in comparison with other environmental pollutants. While toxicity of other contaminants is merely dependent on their composition, that of MPs is more complex. The toxicity of MPs includes the particle-related toxicity, which is driven by size, shape, colour and the polymer type; and the chemical toxicity, which is driven by adsorption-desorption kinetics of additives, and co-existing pollutants [6]. This section reviews the implications of MPs' characteristics in their behaviour and fate, and further highlights the consequences of these implications on their bioavailability and potential toxicity.

4.1 Size

The size of MPs influences their distribution in the environment, dispersal in water column, magnitude of buoyant, biofilm formation and sedimentation. It also determines the extent of their impacts on soil properties, bioavailability, plant growth, GHGs emission rates and their potential health risks [21, 54–56].

For instance, their impact on the climate depends to a great extent on their size. For example, the larger surface area of the small-sized MPs increases the emission rate of GHGs. Meaning, with the frequent degradation of MPs, the very same amount of MPs will continue to release more and more GHGs [23, 24].

In terrestrial environment, small-sized MPs are more likely to block soil micropores, absorb by plants and be consumed by soil organisms compared to larger MPs. In addition, they are transported through the soil to groundwater more easily than larger MPs [57].

In aquatic systems, the size range of MPs overlaps with the preferred particle size ingested by a wide range of aquatic biota, including filters, detritus and suspension feeders. In addition, some organisms such as diatoms can aggregate on the surfaces of small-sized MPs and construct biofilms that could be attractive to organisms causing a higher probability of being mistakenly ingested. Smaller MPs generally have larger surface area, which makes them a good carrier for other pollutants such as heavy metals [5, 58–62].

In terms of toxicity, pollutants' adsorption and release from MPs depend, among other parameters, on the total surface area and thus on the size of the particles. The greater surface area of small-sized MPs thus facilitates the adsorption of other pollutants from the surrounding environment, resulting in additional health risks [63]. Consequently, small-sized MPs are considered to be more harmful to aquatic organisms. Hamed et al. [51] examined the effects of varying sizes of MPs in fish, and they observed toxicological effects including oxidative stress, biochemical changes and immune destruction. These toxicological effects were found to be augmented with decreasing MP size, thus, implying a direct correlation between the toxicity of MPs and their size. Additionally, the small size may facilitate their translocation into other organs. For instance, MPs have been reported in tissues, muscles and organs, confirming their ability to be translocated into these parts of the body, and it was noted that translocation rates increased with decreasing particle size [64–66]. This represents higher potential for health risks and higher level of toxicity if small MPs are regularly translocated into other parts of the body.

4.2 Shape

MPs' shape is a key attribute affecting their behaviour in the environment. It influences biofouling, rising and fall velocities and drag force. Thus, the shape plays a significant role in the sedimentation of MPs [67–69]. The shape of MPs is also important with regards to their impact on soil properties.

Some studies have suggested that MPs with different shapes may affect soil properties differently. For instance, fibres and films may have more significant effects on soil properties compared to beads and spheres [57, 70, 71]. This was explained by Rillig et al. [72] that the pollutants with dissimilar shapes to soil particles may have stronger effects. This was further supported by Lozano et al. [57]; they reported different effects on soil based on different shapes. They observed different effects caused by different shapes; according to their findings, fibres increased water-holding capacity, films decreased soil bulk density, while foams and fragments increased soil aeration and porosity.

In terms of bioavailability and toxicity, the shape of MPs is essential to prey perception by visual predator and the preference for certain MP shapes by several aquatic organisms have been reported in several studies. For instance, Saad et al. [5] and Yuan et al. [73] observed that common carp fish and goldfish preferably consumed fibrous MPs in the presence of other shapes, whereas, Hurley et al. [74] and Schessl et al. [75] reported an absence of pellets in the freshwater worm *Tubifex tubifex* and bivalves *(Dreissena polymorpha and D. bugensis)* despite their presence in the environment. This confirms the role of MPs shape in their bioavailability to different aquatic organisms.

Further, MPs' shape is pertinent to their potential toxic effects due to the different retention time, accumulation and physical damage. For example, fibres are reported to have longer intestinal residence time and accumulation, stronger acute toxicity and intestinal epithelial cell necrosis compared to other shapes in zebrafish, amphipods and grass shrimp. This could be attributed to the non-spherical shape, which is more easily embedded in tissue and takes longer time to pass through the gut. It is known that the longer the particles remain within the organism, the greater the potential to release associated toxins [76–79].

4.3 Surface morphology

The surface morphology of MPs influences their interaction with the surrounding environment. For instance, adsorption/desorption of co-existing pollutants as well as biofilm formation are influenced, to a great extent, by the surface nature of MPs. Ultimately, surface morphology impacts the distribution and sedimentation, thus bioavailability of MPs [80].

Cracks, pitting, flaking and fracturing result in an increased surface area, which increases the emission of GHGs. The increased surface area also facilitates the adsorption of other pollutants as well as the formation of biofilms. As a result, MPs may become a cocktail of pollutants with varying toxicity effects [10, 11, 23, 30, 47, 81].

4.4 Colour

A variety of colourant agents such as pigments and dyes are widely used during plastic manufacturing, these colourants contain some toxic chemicals. Coloured MPs are, therefore, considered to have higher potential health risks compared to non-coloured MPs [82].

The uptake of coloured MPs by aquatic biota is well documented and exposure studies have suggested that aquatic organisms may actively prey on plastic particles that possess similar colours to their natural prey. For instance, the preferential uptake of certain colours (MPs with artificial food-like colours) was reported in common carp (*Cyprinus carpio*). The authors observed an increase in the number of ingested food-like MPs with increasing concentrations of MPs in the water, while no increase in the number of non-food-like colours was observed [83]. Similarly, Ory et al. [84] reported significant uptake of MPs with artificial food-like colours by palm ruff (*Seriolella violacea*) more often than other colours, whereas MPs of other colours were mostly co-ingested when floating close to food pellets. Another study by de Sá et al. [85] reported a preferential uptake of white MPs by common goby (*Pomatoschistus microps*), compared to black and red MPs. The authors attributed this to the similarity in colour with the brine shrimp (*Artemia nauplii*), a prey that is commonly consumed by the common goby.

4.5 Chemical composition

Generally, MPs consist of a high molecular polymer as the main body and a variety of additives such as stabilizers, plasticizers, flame retardants and colouring agents as auxiliary materials. These various potential compositions determine their properties such as density and degradability, behaviour and environmental impact. For instance, their distribution in different environmental compartments is greatly influenced by polymer density. MPs with low density are buoyant, while those of high density tend to sink into sediments [37, 68].

The chemical composition of MPs also influences their interaction with the co-existing contaminants. For instance, the tendency of MPs to adsorb metals depends mainly on the functional groups pendent on the backbone structure of the polymer [86–92]. Ultimately, the chemical composition of MPs shapes their toxicological effects.

It is reported that the effect of MPs on soil properties varies based on the polymer type. The polymer type further influences the degradation of MPs, as a result, MPs may release the contaminants that are adsorbed onto their surfaces. The degradation also increases the emission rate of GHGs by MPs. The rate of GHGs emission also depends on the polymer type, for instance, polyethylene is found to emit higher GHGs compared to a number of other polymers [22, 30, 34, 80, 93].

5. Concluding remarks

The ubiquitous detection of MPs in different environmental compartments has made them a prominent environmental concern. Due to the chemical modification of plastic materials, receiving environments are potentially exposed to a cocktail of pollutants (polymers, leached additives and degradation products). This chapter provided a brief overview of the environmental challenges associated with MPs. The complexity of their ecological impact is discussed in light of their heterogeneous physicochemical characteristics.

Over the past decade, monitoring and ecotoxicological studies have improved our understanding of their nature and potential health risks. However, a better understanding of their long-term effects is needed. Considering that MP pollution is a symptom of human-made environmental change and a valid example of society-nature interaction, their mitigation requires a lot more than technological innovation. For instance, strategies such as reuse, design for recycling, improved waste management, standardized labelling and sustainable consumption are important. However, such strategies are only effective through social action and regulations. Examples include behavioural change campaigns to reduce the use of single-use plastics and policy measures to reduce the use of microbeads.

Funding

The author would like to acknowledge the Royal Society, UK for funding through the Grand Challenges Research Fund (GCRF), [grant number: FLR\R1\201062].

Conflict of interest

The author declares no conflict of interest.

Author details

Dalia Saad
School of Chemistry, Molecular Sciences Institute, University of the Witwatersrand, Johannesburg, South Africa

*Address all correspondence to: dalia.saad@wits.ac.za

References

[1] Weis JS. Improving microplastic research. AIMS Environmental Science. 2019;**6**:326-341

[2] Koelmans B, Kalčíková G. A Scientific Perspective on Microplastics in Nature and Society. Berlin, Germany: SAPEA; 2019

[3] Saad D, Ndlovu M, Ramaremisa G, Tutu H. Microplastics in freshwater environment: The first evaluation in sediment of the Vaal River, South Africa. Heliyon. 2022;**8**:e11118

[4] Kalčíková G. Aquatic vascular plants – A forgotten piece of nature in microplastic research. Environmental Pollution. 2020;**262**:114354

[5] Saad D, Chauke P, Cukrowska E, Richards H, Nikiema J, Chimuka L, et al. First biomonitoring of microplastic pollution in the Vaal river using Carp fish (Cyprinus carpio) "as a bio-indicator". The Science of the Total Environment. 2022;**836**

[6] Scherer C, Weber A, Lambert S, Wagner M. Interactions of microplastics with freshwater biota, Handbook of. Environment and Chemistry. 2018;**58**:153-180

[7] Long M, Moriceau B, Gallinari M, Lambert C, Huvet A, Raffray J, et al. Interactions between microplastics and phytoplankton aggregates: Impact on their respective fates. Marine Chemistry. 2015;**2015**

[8] Bellasi A, Binda G, Pozzi A, Galafassi S, Volta P, Bettinetti R. Microplastic contamination in freshwater environments: A review, focusing on interactions with sediments and benthic organisms. Environments - MDPI. 2020;7

[9] Amelia TSM, Khalik WMAWM, Ong MC, Shao YT, Pan HJ, Bhubalan K. Marine microplastics as vectors of major ocean pollutants and its hazards to the marine ecosystem and humans. Progress in Earth and Planetary Science. 2021;**8**:1-26

[10] Caruso G. Microplastics as vectors of contaminants. Marine Pollution Bulletin. 2019;**146**:921-924

[11] Joo SH, Liang Y, Kim M, Byun J, Choi H. Microplastics with adsorbed contaminants: Mechanisms and treatment. Environmental Challenges. 2021;**3**:100042

[12] Banni M, Cholewí Nska P, Moniuszko H, Wojnarowski K, Pokorny P, Szeligowska N, et al. The occurrence of microplastics and the formation of biofilms by pathogenic and opportunistic bacteria as threats in aquaculture. Public Health. 2022;**19**: 8137

[13] Kellyid JJ, London MG, Mccormick AR, Rojas M, Scott JW, Hoellein TJ. Wastewater treatment alters microbial colonization of microplastics. PLoS ONE. 2021;**16**(1):e0244443

[14] Iqbal R, Khan MT, Bilal H, Aslam MM, Khan IA, Raja S, et al. Microplastics as vectors of environmental contaminants: Interactions in the natural ecosystems. Human and Ecological Risk Assessment: An International Journal. 2022;**28**(9):1022-1042

[15] Grbić J, Helm P, Athey S, Rochman CM. Microplastics entering northwestern Lake Ontario are diverse and linked to urban sources. Water Research. 2020;**174**:115623

[16] Zhao S, Zhu L, Wang T, Li D. Suspended microplastics in the surface water of the Yangtze Estuary System, China: First observations on occurrence, distribution. Marine Pollution Bulletin. 2014;**86**:562-568

[17] Aves AR, Revell LE, Gaw S, Ruffell H, Schuddeboom A, Wotherspoon NE, et al. First evidence of microplastics in Antarctic snow. The Cryosphere. 2022;**16**:2127-2145

[18] Yonkos LT, Friedel EA, Perez-Reyes AC, Ghosal S, Arthur CD. Microplastics in four estuarine rivers in the Chesapeake bay, U.S.A. Environmental Science & Technology. 2014;**48**:14195-14202

[19] Qiu Q, Peng J, Yu X, Chen F, Wang J, Dong F. Occurrence of microplastics in the coastal marine environment: First observation on sediment of China. Marine Pollution Bulletin. 2015;**98**:274-280

[20] Shen M, Huang W, Chen M, Song B, Zeng G, Zhang Y. (Micro)plastic crisis: Un-ignorable contribution to global greenhouse gas emissions and climate change. Journal of Cleaner Production. 2020;**254**:120138

[21] Chen X, Huang G, Gao S, Wu Y. Effects of permafrost degradation on global microplastic cycling under climate change. Journal of Environmental Chemical Engineering. 2021;**9**:106000

[22] Ford HV, Jones NH, Davies AJ, Godley BJ, Jambeck JR, Napper IE, et al. The fundamental links between climate change and marine plastic pollution. The Science of the Total Environment. 2022;**806**

[23] Royer SJ, Ferrón S, Wilson ST, Karl DM. Production of methane and ethylene from plastic in the environment. PLoS One. 2018;**13**:e0200574

[24] Shen M, Huang W, Chen M, Song B, Zeng G, Zhang Y. (Micro)plastic crisis: Un-ignorable contribution to global greenhouse gas emissions and climate change. Journal of Cleaner Production. 2020;**254**:120138

[25] Shen M, Ye S, Zeng G, Zhang Y, Xing L, Tang W, et al. Can microplastics pose a threat to ocean carbon sequestration? Marine Pollution Bulletin. 2020;**150**:110712

[26] Yu H, Fan P, Hou J, Dang Q, Cui D, Xi B, et al. Inhibitory effect of microplastics on soil extracellular enzymatic activities by changing soil properties and direct adsorption: An investigation at the aggregate-fraction level. Environmental Pollution. 2020;**267**:115544

[27] Chia RW, Lee J-Y, Jang J, Kim H, Kwon KD. Soil health and microplastics: A review of the impacts of microplastic contamination on soil properties. Journal of Soils and Sediments. 2022;**22**:2690-2705

[28] Crossman J, Hurley RR, Futter M, Nizzetto L. Transfer and transport of microplastics from biosolids to agricultural soils and the wider environment. The Science of the Total Environment. 2020;**724**

[29] Kim YN, Yoon JH, Kim KH. Microplastic contamination in soil environment - A review. Soil Science Annual. 2020;**71**:300-308

[30] Feng X, Wang Q, Sun Y, Zhang S, Wang F. Microplastics change soil properties, heavy metal availability and bacterial community in a Pb-Zn-contaminated soil. Journal of Hazardous Materials. 2022;**424**:127364

[31] Zhang Z et al. Current research and perspective of microplastics (MPs) in

soils (dusts), rivers (lakes), and marine environments in China. Ecotoxicological and Environmental Safety. 2020;**202**

[32] de Souza MacHado AA, Lau CW, Till J, Kloas W, Lehmann A, Becker R, et al. Impacts of Microplastics on the Soil Biophysical Environment. Environmental Science & Technology. 2018;**52**:9656-9665

[33] Abel De Souza Machado A, Lau CW, Kloas W, Bergmann J, Bachelier JB, Faltin E, et al. Microplastics can change soil properties and affect plant performance. Environmental Science & Technology. 2019;**53**(10):6044-6052

[34] Wang F, Wang Q, Adams CA, Sun Y, Zhang S. Effects of microplastics on soil properties: Current knowledge and future perspectives. Journal of Hazardous Materials. 2022;**424**:127531

[35] Wang J, Liu X, Li Y, Powell T, Wang X, Wang G, et al. Microplastics as contaminants in the soil environment: A mini-review. Science of The Total Environment. 2019;**691**:848-857

[36] Li H-Z, Zhu D, Lindhardt JH, Lin S-M, Ke X, Cui L. Long-term fertilization history alters effects of microplastics on soil properties, microbial communities, and functions in diverse farmland ecosystem. Environmental Science & Technology. 2021;**2021**

[37] Salthammer T. Microplastics and their additives in the indoor environment. Angewandte Chemie, International Edition. 2022;**61**:e202205713

[38] Campanale C, Massarelli C, Savino I, Locaputo V, Uricchio VF. A detailed review study on potential effects of microplastics and additives of concern on human health. International Journal of Environmental Research and Public Health. 2020;**17**

[39] Ramaremisa G, Ndlovu M, Saad D. Comparative assessment of microplastics in surface waters and sediments of the Vaal River, South Africa: Abundance, composition, and sources. Environmental Toxicology and Chemistry. 2022;**41**:3029-3040

[40] Wang J, Guo X, Xue J. Biofilm-developed microplastics as vectors of pollutants in aquatic environments. Environmental Science & Technology. 2021;**55**:12780-12790

[41] Brennecke D, Duarte B, Paiva F, Caçador I, Canning-Clode J. Microplastics as vector for heavy metal contamination from the marine environment. Estuarine, Coastal and Shelf Science. 2016;**178**:189-195

[42] Hartmann NB, Rist S, Bodin J, Jensen LHS, Schmidt SN, Mayer P, et al. Microplastics as vectors for environmental contaminants: Exploring sorption, desorption, and transfer to biota. Integrated Environmental Assessment and Management. 2017;**13**:488-493

[43] Fred-Ahmadu OH, Bhagwat G, Oluyoye I, Benson NU, Ayejuyo OO, Palanisami T. Interaction of chemical contaminants with microplastics: Principles and perspectives. Science of The Total Environment. 2020;**706**:135978

[44] Santos LHMLM, Rodríguez-Mozaz S, Barceló D. Microplastics as vectors of pharmaceuticals in aquatic organisms – An overview of their environmental implications. Case Studies in Chemical and Environmental Engineering. 2021;**3**:100079

[45] Hanslik L, Huppertsberg S, Kämmer N, Knepper TP, Braunbeck T.

Rethinking the relevance of microplastics as vector for anthropogenic contaminants: Adsorption of toxicants to microplastics during exposure in a highly polluted stream - Analytical quantification and assessment of toxic effects in zebrafish (Danio rerio). Science of The Total Environment. 2022;**816**:151640

[46] Naqash N, Prakash S, Kapoor D, Singh R. Interaction of freshwater microplastics with biota and heavy metals: A review. Environmental Chemistry Letters 2020 18. 2020;**6**:1813-1824

[47] Banni M, Cholewí Nska P, Moniuszko H, Wojnarowski K, Pokorny P, Szeligowska N, et al. The occurrence of microplastics and the formation of biofilms by pathogenic and opportunistic bacteria as threats in aquaculture. Public Health. 2022;**19**:8137

[48] Chen J, Rao C, Yuan R, Sun D, Guo S, Li L, et al. Long-term exposure to polyethylene microplastics and glyphosate interferes with the behavior, intestinal microbial homeostasis, and metabolites of the common carp (*Cyprinus carpio* L.). Science of The Total Environment. 2022;**814**:152681

[49] Xia X, Sun M, Zhou M, Chang Z, Li L. Polyvinyl chloride microplastics induce growth inhibition and oxidative stress in Cyprinus carpio var. larvae. Science of The Total Environment. 2020;**716**:136479

[50] Rochman CM, Hoh E, Kurobe T, Teh SJ. Ingested plastic transfers hazardous chemicals to fish and induces hepatic stress. Scientific Reports. 2013;**3**:1-7

[51] Hamed M, Monteiro CE, Sayed AEDH. Investigation of the impact caused by different sizes of polyethylene plastics (nano, micro, and macro) in common carp juveniles, Cyprinus carpio L., using multi-biomarkers. Science of The Total Environment. 2022;**803**:149921

[52] Nematdoost Haghi B, Banaee M. Effects of micro-plastic particles on paraquat toxicity to common carp (Cyprinus carpio): Biochemical changes. International Journal of Environmental Science and Technology. 2017;**14**:521-530

[53] Hatami M, Banaee M, Nematdoost Haghi B. Sub-lethal toxicity of chlorpyrifos alone and in combination with polyethylene glycol to common carp (Cyprinus carpio). Chemosphere. 2019;**219**:981-988

[54] Shamskhany A, Li Z, Patel P, Karimpour S. Evidence of microplastic size impact on mobility and transport in the marine environment: A review and synthesis of recent research. Frontiers in Marine Science. 2021;**8**:1869

[55] Zhang X, Chen Y, Li X, Zhang Y, Gao W, Jiang J, et al. Size/shape-dependent migration of microplastics in agricultural soil under simulative and natural rainfall. Science of The Total Environment. 2022;**815**:152507

[56] Coverntton GA, Pearce CM, Gurney-Smith HJ, Chastain SG, Ross PS, Dower JF, et al. Size and shape matter: A preliminary analysis of microplastic sampling technique in seawater studies with implications for ecological risk assessment. Science of The Total Environment. 2019;**667**:124-132

[57] Lozano YM, Lehnert T, Linck LT, Lehmann A, Rillig MC. Microplastic shape, polymer type, and concentration affect soil properties and plant biomass. Frontiers in Plant Science. 2021;**12**:169

[58] Wang W, Ge J, Yu X. Bioavailability and toxicity of microplastics to fish

species: A review. Ecotoxicology and Environmental Safety. 2020;**189**

[59] Chakraborty P, Chandra S, et al. Interlinkage between persistent organic pollutants and plastic in the waste management system of India: An overview. Bulletin of Environmental Contamination and Toxicology. 2022;**109**(6):927-936

[60] Batel A, Linti F, Scherer M, Erdinger L, Braunbeck T. Transfer of benzo[a]pyrene from microplastics to Artemia nauplii and further to zebrafish via a trophic food web experiment: CYP1A induction and visual tracking of persistent organic pollutants. Environmental Toxicology and Chemistry. 2016;**35**:1656-1666

[61] Rogers KL, Carreres-Calabuig JA, Gorokhova E, Posth NR. Micro-by-micro interactions: How microorganisms influence the fate of marine microplastics. Limnol Oceanogr Lett. 2020;**5**:18-36

[62] Galloway TS, Lewis CN. Marine microplastics spell big problems for future generations. PNAS. 2022;**113**(9):2331-2333

[63] Caputo F, Vogel R, Savage J, Vella G, Law A, et al. Measuring particle size distribution and mass concentration of nanoplastics and microplastics: Addressing some analytical challenges in the sub-micron size range. Journal of Colloid and Interface Science. 2021;**588**:401-417

[64] Collard F, Gasperi J, Gilbert B, Eppe G, Azimi S, Rocher V, et al. Anthropogenic particles in the stomach contents and liver of the freshwater fish Squalius cephalus. Science of The Total Environment. 2018;**643**:1257-1264

[65] Browne MA, Dissanayake A, Galloway TS, Lowe DM, Thompson RC. Ingested microscopic plastic translocates to the circulatory system of the mussel, Mytilus edulis (L.). Environmental Science & Technology. 2008;**42**:5026-5031

[66] Galafassi S, Campanale C, Massarelli C, Uricchio VF, Volta P. Do freshwater fish eat microplastics? A review with a focus on effects on fish health and predictive traits of mps ingestion. Water (Switzerland). 2021;**13**(16):2214

[67] Kowalski N, Reichardt AM, Waniek JJ. Sinking rates of microplastics and potential implications of their alteration by physical, biological, and chemical factors. Marine Pollution Bulletin. 2016;**109**:310-319

[68] Kooi M, van Nes EH, Scheffer M, Koelmans AA. Ups and downs in the ocean: Effects of biofouling on vertical transport of microplastics. Environmental Science & Technology. 2017;**51**:7971

[69] Khatmullina L, Isachenko I. Settling velocity of microplastic particles of regular shapes. Marine Pollution Bulletin. 2017;**114**:871-880

[70] Lehmann A, Leifheit EF, Gerdawischke M, Rillig MC. Microplastics have shape- and polymer-dependent effects on soil aggregation and organic matter loss – an experimental and meta-analytical approach. Microplastics and Nanoplastics. 2021;**1**:1-14

[71] Zhao T, Lozano YM, Rillig MC. Microplastics increase soil pH and decrease microbial activities as a function of microplastic shape, polymer type, and exposure time. Frontiers in Environmental Science. 2021;**9**

[72] Rillig MC, Lehmann A, de Souza Machado AA, Yang G. Microplastic

effects on plants. The New Phytologist. 2019;**223**:1066-1070

[73] Yuan W, Liu X, Wang W, Di M, Wang J. Microplastic abundance, distribution and composition in water, sediments, and wild fish from Poyang Lake, China. Ecotoxicology and Environmental Safety. 2019;**170**:180-187

[74] Hurley RR, Woodward JC, Rothwell JJ. Ingestion of microplastics by freshwater Tubifex worms. 2017;**51**(21):12844-12851

[75] Schessl M, Johns C. Ashpole, microbeads in sediment, Dreissenid Mussels, and Anurans in the littoral zone of the Upper St. Lawrence River, New York. Pollution. 2019;**5**:41-52

[76] Blarer P, Burkhardt-Holm P. Microplastics affect assimilation efficiency in the freshwater amphipod Gammarus fossarum, Environmental Science and Pollution. Research. 2016;**23**:23522-23532

[77] Gray AD, Weinstein JE. Size- and shape-dependent effects of microplastic particles on adult daggerblade grass shrimp (Palaemonetes pugio). Environmental Toxicology and Chemistry. 2017;**36**:3074-3080

[78] Qiao R, Deng Y, Zhang S, Wolosker MB, Zhu Q, Ren H, et al. Accumulation of different shapes of microplastics initiates intestinal injury and gut microbiota dysbiosis in the gut of zebrafish. Chemosphere. 2019;**236**:124334

[79] Yuan Z, Nag R, Cummins E. Human health concerns regarding microplastics in the aquatic environment - From marine to food systems. The Science of the Total Environment. 2022;**823**

[80] Zhang J, Wang L, Kannan K. Quantitative analysis of polyethylene terephthalate and polycarbonate microplastics in sediment collected from South Korea, Japan and the USA. Chemosphere. 2021;**279**

[81] Yu F, Yang C, Zhu Z, Bai X, Ma J. Adsorption behavior of organic pollutants and metals on micro/nanoplastics in the aquatic environment. The Science of the Total Environment. 2019;**694**

[82] Onoja S, Nel HA, Abdallah MAE, Harrad S. Microplastics in freshwater sediments: Analytical methods, temporal trends, and risk of associated organophosphate esters as exemplar plastics additives. Environmental Research. 2022;**203**

[83] Roch S, Friedrich C, Brinker A. Uptake routes of microplastics in fishes: Practical and theoretical approaches to test existing theories. Scientific Reports. 2020;**10**

[84] Ory NC, Gallardo C, Lenz M, Thiel M. Capture, swallowing, and egestion of microplastics by a planktivorous juvenile fish. Environmental Pollution. 2018;**240**:566-573

[85] de Sá LC, Luís LG, Guilhermino L. Effects of microplastics on juveniles of the common goby (Pomatoschistus microps): Confusion with prey, reduction of the predatory performance and efficiency, and possible influence of developmental conditions. Environmental Pollution. 2015;**196**:359-362

[86] Saad DM, Cukrowska EM, Tutu H. Development and application of cross-linked polyethylenimine for trace metal and metalloid removal from mining and industrial wastewaters. Toxicological and Environmental Chemistry. 2011;**93**

[87] Saad D. Selective removal of Cr(III) from aqueous solution using cross-linked

polyethylenimine: Experimental optimization and modeling. Polymer Bulletin. 2022;**79**

[88] Saad DMG, Cukrowska EM, Tutu H. Phosphonated cross-linked polyethylenimine for selective removal of uranium ions from aqueous solutions. Water Science and Technology. 2012;**66**

[89] Saad DMG, Cukrowska EM, Tutu H. Sulfonated cross-linked polyethylenimine for selective removal of mercury from aqueous solutions. Toxicological and Environmental Chemistry. 2012;**94**

[90] Saad DM, Cukrowska EM, Tutu H. Selective removal of mercury from aqueous solutions using thiolated cross-linked polyethylenimine. Applied Water Science. 2013;**3**

[91] Saad DMG, Cukrowska EM, Tutu H. Modified cross-linked polyethylenimine for the removal of selenite from mining wastewaters. Toxicological and Environmental Chemistry. 2013;**95**

[92] Saad DM, Cukrowska EM, Tutu H. Functionalisation of cross-linked polyethylenimine for the removal of as from mining wastewater. Water SA. 2013;**39**

[93] Selonen S, Dolar A, Jemec Kokalj A, Skalar T, Parramon Dolcet L, Hurley R, et al. Exploring the impacts of plastics in soil – The effects of polyester textile fibers on soil invertebrates. Science of The Total Environment. 2020;**700**:134451

Chapter 2

Bioplastics against Microplastics: Screening of Environmental Bacteria for Bioplastics Production

Aisha S. Alwuhaib, Vitaly Zinkevich, Tamar Kartvelishvili, Nino Asatiani and Nelly Sapojnikova

Abstract

Polyhydroxyalkanoates (PHAs) are biopolymers produced by numerous bacteria and can be used in the production of bioplastics. PHAs are synthesized by microorganisms by fermentation of carbon sources. Due to the different monomer structures of PHAs, there are many kinds of PHAs, and their corresponding material properties are also very different. Thus, the search for bacteria producing the PHAs is of great interest. In this study, the bacteria isolated from the environment were analyzed for the presence of PHA. PHA production was tested with staining methods Sudan Black B, Nile Blue, and Nile Red. The presence of a PHA synthase gene (*phaC*) was confirmed by PCR amplification. PHAs were extracted from the strains and characterized by the FTIR spectroscopy method. A biochip for a fast screening of environmental samples for the presence of PHA-producing bacteria was designed. The biochip contained 11 probes for coding class 1, 2, and 3 PHA synthase genes.

Keywords: polyhydroxyalkanoates (PHAs), PHA synthase gene, environmental bacteria, biochip, bioplastics

1. Introduction

Plastics (polymeric materials) are highly functional materials that have become an essential part of the products we use in our daily life due to their ease of production and robustness. Around 360 million tons of plastics have been produced in 2020 around the world [1]. As more plastics are used, especially with their short-use life span, more waste surfaces. Plastic trash photodegrades into smaller fragments (microplastics). As it was emphasized in Ref. [1], the worldwide use of disposable face masks during the pandemic time and still now is an additional source of microplastics in the environment. Plastics are majorly manufactured from petrochemical feedstock, accounting for 80% of the total produced plastics. The building blocks for the polymerization of bioplastics are biopolymers cultivated from renewable production pathways, such as polymers from microorganisms [2, 3]. Bio-based polymers have a lower carbon footprint than petrochemicals because they utilize biological materials and waste, which convert biocaptured CO_2 into durable polymeric materials. It is a well-established fact that microorganisms are equipped with diverse metabolic activities that enable them to work as biorefineries for transforming a wide range of petrochemical and organic wastes into high-value specific products, such as polyhydroxyalkanoates biopolymers, while positively impacting the carbon cycle by consuming atmospheric CO_2 [4]. The chemical structure of PHA is shown in **Figure 1**. PHA is a linear polyester that contains 3-hydroxy fatty acid monomers [5]. The most commonly produced PHA is poly 3-hydroxybutyrate (PHB), where the alkyl group is R = CH_3. However, there are over 150 different monomers, such as polyhydroxyvalerate (PHV, R = C_2H_5), polyhydroxyhexanoate (PHH, R = C_3H_7), and polyhydroxyoctanote (PHO, R = C_4H_9). PHAs are high molecular weight linear polyesters, ranging in size from 50 to 10,000 kDa [2], characterized by a diversity of structures defined by the length of the carbon chain, referred to as short-chain length (SCL), medium-chain length (MCL) [6], and long-chain length (LCL) [7] PHAs.

A very important property of biopolymers is hydrophobicity, which determines their solubility, biodegradability, and biocompatibility. As PHAs contain chains of hydrophobic groups with different lengths and structures, and at the same time PHAs are poorly hydrophilic due to the presence of carbonyl groups, the hydrophobicity-hydrophilicity balance is a very important point in the selection of appropriate bacteria that give the desired biomaterials [2]. PHAs can exist as homo or copolymers. PHB is a homopolymer, which has a linear isotactic structure, that is, highly crystalline making it brittle and unsuitable for many applications. This problem can be circumvented by forming a copolymer. The first commercially manufactured PHA, Biopol®, is copolymer produced from poly (3-hydroxybutyrate-co-3-hydroxyvalerate (PHB/PHV) that had an increased side chain length making it less crystalline and more ductile [8]. These biopolymers have been produced commercially since the 1980s, and are currently marketed as Mirel®. The combinations in different proportions of the available PHAs create copolymer plastics with various properties [9]. PHAs are synthesized by various types of bacteria in the form of water-insoluble granules, and are stored as carbonosomes within the cell cytoplasm [10]. Bacteria capable of producing PHAs include species of *Alcaligenes*, *Bacillus*, *Burkholderia*, *Ralstonia*, *Pseudomonas*, *Aeromonas,* and *Thiococcus*, among others [11, 12]. PHA synthases and depolymerases, which are the catalyst enzymes for PHA production and degradation, respectively, are located at the membrane of the storage organelles and control the amount of PHA stored by the cell. There are eight major pathways for PHA synthesis. The main metabolite, acetyl-CoA, provides the various lengths of 3-hydroxyalkanoyl-CoA, which act as substrates for PHA production [13]. These pathways are intricately linked to central metabolic pathways

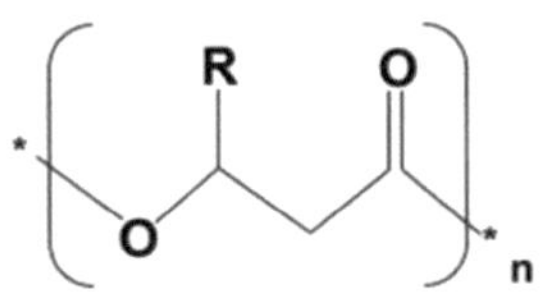

R group	Chemical Name
Hydrogen	Poly-3-hydroxypropionate
Methyl	Poly-3-hydroxybutyrate
Ethyl	Poly-3-hydroxyvalerate
Propyl	Poly-3-hydroxyhexanoate
Pentyl	Poly-3-hydroxyoctanoate

Figure 1.
Chemical structure of polyhydroxyalkanoates (PHAs).

via glycolysis, Kreb's cycle, β-oxidation, fatty acid synthesis, and others. Many enzymes and genes are involved in PHA synthesis. Bacterial cells normally grow when carbon (i.e., fructose and glucose) and nitrogen (ammonium) are present in the medium/environment. Polyhydroxybutyrate (PHB) is produced when the nitrogen source is depleted and carbon is present and will be utilized after refeeding cells with a nitrogen source [14].

The limiting step in the commercial development of non-petrochemical-based produced plastics is the biopolymer yield obtained after fermentation and the cost of its recovery. As the main drawback of bioplastics is the cost of the fermentation process, this has led to searches for cheaper carbon sources for fermentation, which has included activated sludge, paper and food wastes, wastewater, and various oil wastes [15]. Additionally, a lot of effort has been directed toward isolating PHA-producing bacteria from high carbon, low nitrogen, or phosphorus environments that might give greater yields in batch or continuous cultures [16]. The use of different carbon sources in fermentations and isolating producing strains from carbon-rich environments has led to the discovery of novel PHA polymers with different properties. In this study, the results of screening of environmental samples for PHA-producing organisms, isolation and characterization of the microorganisms, detection of genes, coding for enzymes involved in PHA synthesis, and physical characterization of the produced PHAs biopolymers are presented.

2. Analysis of single bacterial isolates for PHA production using staining methods

Two types of samples were collected for this study. The sludge samples were collected from the Petersfield Southern Water Treatment plant (Hampshire, UK). The three compost samples were collected from (1) a graveyard, (2) food-based waste and (3) garden waste (Portsmouth, UK). A mass of 0.5 g of sludge and compost was added to the 50 ml of the mineral salt medium (MSM) and Lysogeny broth (LB) medium (Fisher BioReagents, UK); each sample was inoculated in duplicate for each media and was transferred to the Laboratory of Molecular Microbiology (University of Portsmouth, UK). These bottles were incubated overnight in Innova™ 4000 incubator shaker (New Brunswick Scientific, USA) at 150 rpm at 25°C. Overnight-grown cultures were used for the isolation of single bacterial colonies. Serial dilutions were made into six concentrations using 0.9% NaCl in dH_2O for each culture after incubation. Six dilution tubes were generated for each media (two types of media) and sample (two types of samples), resulting in a total of 48 tubes. Each individual sample from each of the 48 tubes was used to inoculate one LB agar Petri dish. Some of the single colonies that grew from the two most diluted cultures were collected. The collected single colonies were regrown on separate LB agar to ensure that each Petri dish contains a pure colony. The colonies were picked up from Petri dishes and inoculated in individual universal tubes with 5 ml LB, rotating at 37°C overnight at 150 rpm. As a result, 73 bacterial isolates were recovered.

Different approaches have been used for the screening of PHA-producing microbes and/or the imaging of PHA granules [17]. The methods most widely used for detecting PHAs are staining techniques using Nile Red [18], Nile Blue A [19], and Sudan Black B [20]. Due to the lipophilicity of the dyes, the tests are very useful, but they have the ability to nonspecifically bind to other lipid droplets, membranes, and cell envelopes, leading to an incorrect answer [21, 22]. Sudan Black B (SBB) is a diazo fat-soluble dye. PHAs are observed as black granules with SBB by bright-field microscopy. Nile Blue A and Nile Red are highly fluorescent organic dyes belonging to the benzophenoxazine family. Nile Red is a neutral molecule that is poorly soluble in water, and its chromophore is highly susceptible to changes in solvent polarity, showing little or no fluorescence in most popular solvents. Nile Blue A is a cationic dye and is thus more soluble in water than Nile Red. Both fluorescent dyes are particularly useful for visualizing hydrophobic cell structures, such as membranes or lipid-like inclusions (PHA granules). PHA inclusions appear as brightly fluorescent red/orange granules with Nile Blue A and Nile Red [23].

In this study, 73 bacterial isolates were tested for PHAs production using all above mentioned staining methods. When the ability of isolates to produce PHA was screened by Sudan Black B staining, cultures showed granules filled up with dark staining, as PHA granules can be observed as bluish dark spots under a light microscope (**Figure 2**). Out of the tested 73 isolates, 48 isolates tested positive for SBB staining and were considered to produce PHA granules.

To test the possible presence of PHAs additional stains were used. The Sudan Black B-positive strains were stained by Nile Blue A and Nile Red. Both of these dyes are used to detect PHAs production in bacteria grown on solid media. Bacteria were grown on the LB agar plates in the presence of DMSO, incubated at 37°C overnight, and then exposed to UV light to check the fluorescence in the cells (**Figure 3**).

Eighteen colonies out of 73 showed bright orange fluorescence by both fluorescent dye staining. Unfortunately, the staining methods cannot give an unambiguous answer. Therefore, other testing methods were essential to confirm the presence of PHAs in bacteria on the genetic and structural levels.

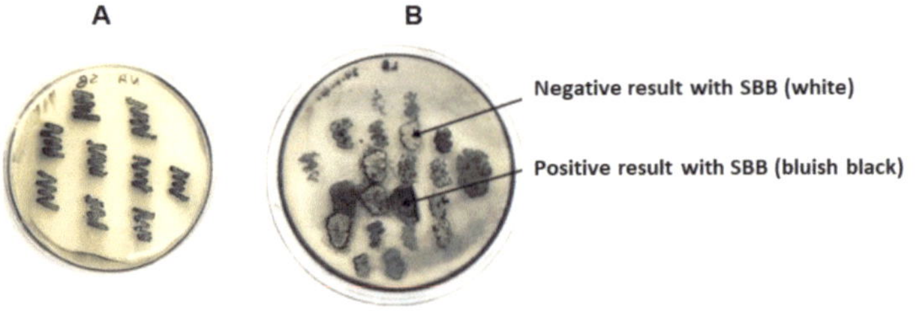

Figure 2.
Screening of bacterial strains for PHA production using Sudan Black B dye. Panel A shows the positive result with different isolates from compost samples stained with SBB. Panel B shows the positive and negative results for different isolates from the sludge sample stained with SBB.

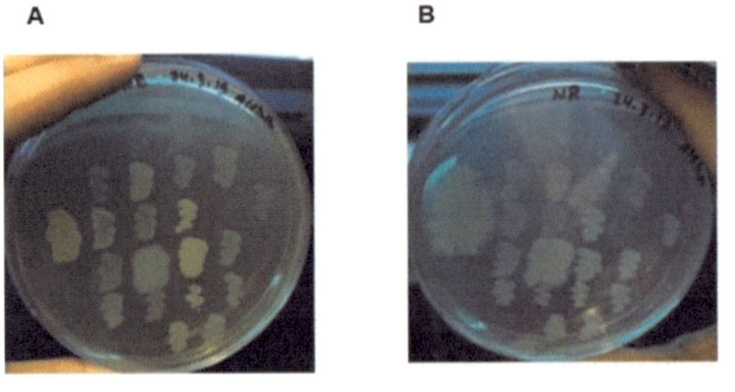

Figure 3.
Screening of different isolates for PHA production using Nile Blue A dye (Panel A) and Nile Red dye (Panel B).

3. PCR amplification of PHAs genes

The genomic DNA from the 48 strains that tested positive in the SBB staining test was analyzed for the presence of the PHA synthase gene (phaC) using the PCR amplification method. Genomic DNA was recovered using the DNeasy Blood and Tissue kit (Qiagen®) according to the manufacturer's instructions. The phaC gene was amplified using F I-179 L forward primer and R I-179R reverse primer (**Table 1**) [24]. Final reactions contained: 2x GoTaq® Green Master Mix and 10 μM of each oligonucleotide primer. PCR was carried out as indicated: initially denatured at 94°C for 10 min; followed by 35 cycles of 94°C for 1 min, 53°C for 1 min, and 72°C for 1 min followed by the final extension step at 72°C for 10 min.

The phaC gene was also amplified using the PhaCF1BO forward primer and PhaCR2BO reverse primer (**Table 1**) [25]. Final reactions contained: 2x GoTaq® Green Master Mix and 10 μM of each oligonucleotide primer. PCR was carried out as indicated: initially denatured at 95°C for 10 min; followed by 30 cycles of 95°C for 1 min, 57°C for 45 sec, and 72°C for 2 min followed by the final extension step at 72°C for 5 min.

The selective results of the amplification using the pair of primers F I-179 L/R I-179R are presented in **Figure 4**. The isolates tested positive for the presence of the

Primer	Sequences of primers 5′ → 3′	Gene target	Size of amplicon, bp
8F	AGAGTTTGATCCTGGCTCAG	*16 S rRNA*	1500
1492R	GGTTACCTTGTTACGACTT		
F I-179L	ACAGATCAACAAGTTCTACATCTTCGAC	*phaC & phbC*	540
R I-179R	GGTGTTGTCGTTGTTCCAGTAGAGGATGTC		
PhaCF1BO	TCMYCTSKACTGCSCTGGYG	*phaC*	247
PhaCR2BO	YWGCTRGACYAGACCTGGAT		

Table 1.
Oligonucleotide primers are used in PCR.

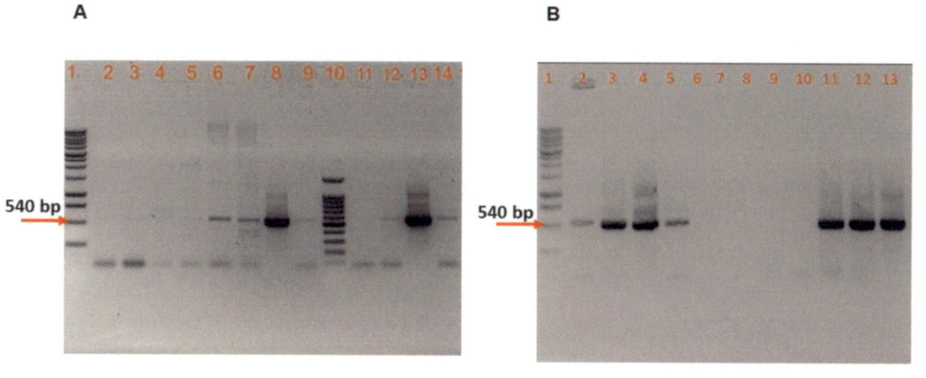

Figure 4.
The DNA amplicons of phaC gene (540 bp) with primers: F I179L/ R I179R from the bacterial isolates. Panel A: Lane 1: 1 kb DNA Ladder; Lane 2: 3/1; Lane 3: 10/1; Lane 4: 12/1; Lane 5: 17/1; Lane 6: 19/1; Lane 7: 20/1; Lane 8: 22/1; Lane 9: 25/1; Lane 10: 100 bp DNA Ladder; Lane 11: 9/1; Lane 12: 6/2; Lane 13: 8/2; Lane 14:10/2; Panel B: Lane 1: 1 kb DNA Ladder. The isolates presented in the Lanes 2 to 13 are as follows: Lane 2: 13/2; Lane 3: 16/2; Lane 4: 17/2; Lane 5: 18/2; Lane 6 to 9 negative results for isolates 2/1, 4/1, 3/2, 14/2; Lane 10: 9/2; Lane 11: 19/2; Lane 12: 24-1/2; Lane 13: 24/2.

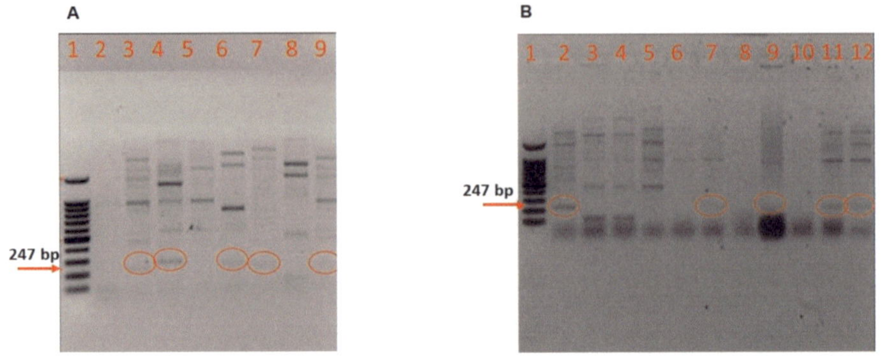

Figure 5.
The DNA amplicons of phaC gene (247 bp) with primers: PhaCF1BO/PhaCR2BO from the bacterial isolates. Panel A: Lane 1: 1 kb DNA Ladder; Lane 2: 3/1; Lane 3: 10/1; Lane 4: 12/1; Lane 5: 17/1; Lane 6: 19/1; Lane 7: 20/1; Lane 8: 22/1; Lane 9: 25/1. Panel B: Lane 1: 1 kb DNA Ladder. The isolates presented in the Lanes 2 to 12 are as follows: Lane 2: 9/1; Lane 3: 6/2; Lane 4: 8/2; Lane 5: 10/2; Lane 6: 3/1; Lane 7: 10/1; Lane 8: 12/1; Lane 9: 17/1; Lane 10: 19/1; Lane 11: 20/1; Lane 12: 22/1.

phaC gene in case in their DNA correct amplicon size (540 bp) was revealed when run on the 1.2% agarose gel (**Figure 4A** and **B**).

The selective results of the amplification using the pair of primers PhaCF1BO/PhaCR2BO are presented in **Figure 5**. According to **Figure 5**, some of the isolates tested positive for the presence of the *phaC* gene in their DNA as shown in lanes (3, 4, 6, 7, and 9) (**Figure 5A**) and lanes (2, 7, 9, 11, and 12) (**Figure 5B**). These isolates produce the correct amplicon size (247 bp) when run on the 1.2% agarose gel.

Pseudomonas oleovorans NCTC 10692 (Bacteria Collection from Public Health England, ATCC 8062), which is the producer of PHA [26, 27] was used as a positive-control strain. *Escherichia coli* XL-1 Blue (Agilent Technologies), which is not able to produce PHA, was used as a negative control strain.

Out of the tested 48 strains, 18 strains tested positive for the presence of the *phaC* genes confirmed by one or both types of PCR and by the presence of PHA granules confirmed by the Nile Blue A and Nile Red staining methods. These 18 strains, in addition to two strains (9/1 and 9/2) that tested negative with PCR and positive with Sudan Black B were sent for sequencing after cloning their 16S rRNA PCR products into competent *E. coli* cells (**Table 2**). The 16S rRNA gene was amplified using the primer pair 8F and 1492R (**Table 1**) [28], the conditions were pointed in Ref. [29].

Sample No.	Strains	Presence of *phaC* gene, PCR	Sudan Black B	Nile Blue A and Nile Red	Presence of plasmid	NCBI BLAST matches (Strain identification name)	Identity, %
1	3/1	+	+	+	+	*Citrobacter sp.*	97
2	9/1	–	+	–	+	*Arthrobacter sp.*	97
3	10/1	+	+	+	+	*Raoultella/Klebsiella*	99
4	12/1	+	+	+	+	*Raoultella/Klebsiella*	98
5	17/1	+	+	+	+	*Alcaligenaceae; Achromobacter*	97
6	19/1	+	+	+	+	*Citrobacter sp.*	99
7	20/1	+	+	+	+	*Raoultella/Klebsiella*	99

Sample No.	Strains	Presence of *phaC* gene, PCR	Sudan Black B	Nile Blue A and Nile Red	Presence of plasmid	NCBI BLAST matches (Strain identification name)	Identity, %
8	22/1	+	+	+	+	*Raoultella/Klebsiella*	99
9	25/1	+	+	+	+	*Serratia*	98
10	6/2	+	+	+	+	*Aeromonas*	99
11	8/2	+	+	+	+	*Alcaligenes*	98
12	9/2	–	+	–	+	*Arthrobacter*	99
13	10/2	+	+	+	+	*Bacillus*	99
14	13/2	+	+	+	+	*Bacillus*	99
15	16/2	+	+	+	+	*Bacillus*	99
16	17/2	+	+	+	+	*Alcaligenes*	99
17	18/2	+	+	+	+	*Alcaligenes*	99
18	19/2	+	+	+	+	*Acinetobacter*	99
19	24-1/2	+	+	+	+	*Pseudomonas*	99
20	24/2	+	+	+	+	*Pseudomonas*	99

Note: (−) indicates a negative test result; and (+) indicates a positive test result.

Table 2.
The characteristics and sequence identification of the selected bacterial isolates from the environmental samples.

PCR products were ligated into a vector plasmid DNA and transformed into *E. coli* competent cells using the TOPO® TA Cloning® Kit (Invitrogen by Life Technologies, UK) following the manufacturer's protocol. Sequencing was performed by GATC Biotech Ltd., London Bioscience Innovation Center, the identification was done through the BLAST search. The strains, which were isolated from sludge and compost samples, belonged to multiple bacterial families (**Table 2**), including the *Enterobacteriaceae* (7 strains, 35%), *Micrococcaceae* (2 strains, 10%), *Alcaligenaceae* (4 strains, 20%), *Moraxellaceae* (1 strain, 5%), *Aeromonadaceae* (1 strain, 5%), *Bacillaceae* (3 strains, 15%), and *Pseudomonadaceae* (2 strains, 10%). Most of the isolates from the environmental samples belong to *Enterobacteriaceae* (*Citrobacter* sp., *Raoultella*, *Klebsiella*, *Serratia*, and *Acinetobacter*), and they are commonly known as producers of polyhydroxyalkanoates [30]. The genomes of the identified isolates contain different classes of PHA synthase (I, II, and IV), producing different types of PHA based on the number of carbon atoms. Thus, *Pseudomonas* contains either class I or II of PHA synthase [12] and can produce medium-chain length PHA, but some strains are able to synthesize both SCL- and MCL-PHAs; *Alcaligenes* can produce short-chain length PHA.

4. Confirmation of the presence of a PHA pathway

The isolates were screened for the presence of plasmids, as they often carry the PHA biosynthetic pathway. Plasmids were detected in all 20 selected strains for PHA production (**Table 2**). To show that the plasmid and PHA production were co-inherited, a non-PHA-producing strain of *E. coli* XL-1 Blue (Agilent Technologies) was transformed with a plasmid. This will allow the identification of plasmids that can express the *PHA* operon in *E. coli*. As a control, a known PHA-producing plasmid from *Pseudomonas oleovorans* NCTC 10692 was transformed to show that the insertion was successful. The plasmid DNA was purified from the bacterial isolates grown in culture in the LB media using the Zyppy miniprep kit (ZYMO Research, Irvine, CA) according to the manufacturer's protocol. Purified plasmid DNA was used for the transformation of *E. coli* XL-1 Blue by electroporation (MicroPulser Electroporator, BioRad). Before performing the transformation experiments, the PHA-producing strains were characterized for their antibiotic resistance profiles. This was done to identify possible determinants that could be used as counter-selective agents, assuming that it was plasmid bourn. The 20 strains were tested for their susceptibility to 16 different antibiotics, delivered in discs to check the resistance profiles of these strains grown on solid media. Four of the PHA-producing strains (6/2, 8/2, 10/2, 13/2) were sensitive to all the tested antibiotics. As all the PHA-producing strains were resistant to ampicillin and/or nalidixic acid, but the *E. coli* XL-1 Blue host strain was not, these two antibiotics were then used as a selection for the plasmid when it was transformed into the PHA-negative strain (*E. coli* XL-1 Blue). After the electroporation, transformed cells were inoculated on LB agar containing ampicillin or nalidixic acid. After overnight incubation, the colonies obtained were picked up and grown in LB medium with the appropriate antibiotics chosen (ampicillin and nalidixic acid). These cultures were then screened for PHA-producing genes by using the staining methods, PCR amplification, and sequencing. Four strains (3/1, 10/1, 17/1, and 19/1) in addition to the positive-control strain were successfully transferred into *E. coli* XL-1 Blue, and were tested positive for PHA production and the presence of the *phaC* gene. These four strains were isolated from the sludge samples collected from the Petersfield Southern Water Treatment plant. The pointed bacterial strains provide plasmids, which have the potential to synthesize PHA for industrial/commercial purposes when co-introduced into the same genetic background.

5. Physical characteristics of PHAs

The PHA extracted from the 20 environmental samples that tested positive for *phaC* genes and/or the staining methods were analyzed using the FTIR spectroscopy method.

Pure bacterial overnight culture grown in LB media (10 ml) at 37°C was used for the PHA extraction and purification. Bacterial cells were pelleted at 4000 rpm for 25 min at RT using glass conical centrifuge tubes (Sigma, UK). The pellet was suspended in 10 ml of 0.1% sodium hypochlorite and centrifuged for 30 min at 4000 rpm. The supernatant was removed, and the pellet was washed with 5 ml of sterile water, 5 ml of acetone, and 5 ml of methanol consequently. The pellet was dissolved in 5 ml of chloroform and left overnight at RT to evaporate on a glass Petri dish (**Figure 6**) [31]. The PHA biopolymer was collected, weighed, and analyzed using a physical method.

The PHAs extracted from the isolates were analyzed using PerkinElmer®'s Spectrum 1™ FT-IR Spectrometer (USA); spectral range 4000–400 cm^{-1} with a spectral resolution of 4 cm^{-1} [32]. The characteristic absorption peaks were used to interpret the presence of specific functional groups in the extracted polymers. The yield of PHA in mg/10 ml extracted from the environmental pure bacterial isolates varied from 0.1 mg (24/2) to 8.8 mg (20/1). The main FTIR spectral peaks for the presumptive

Figure 6.
PHA was obtained as a white powdery substance on a glass Petri dish.

PHAs extracted from 20 studied bacterial isolates; from the PHA-producing bacterial strain (+ve), used as a positive control, along with peaks for the commercial PHB (Sigma, USA) are presented in **Table 3**.

According to **Table 3**, the spectra of the standard PHB show the peak at 1720.88 cm^{-1}, which corresponds to the C=O stretch of the ester group, and the peak at 1278.59 cm^{-1}, which corresponds to –CH group [33]. These peaks are similar to the published spectra peaks of PHB [34]. The positive PHA production bacterial strain (+ve) showed the peak at 1650.05 cm^{-1}, which corresponds to ester carbonyl group (C–O), and accompanying peak at 1275.33 cm^{-1}, which corresponds to –CH group.

Strain ID	Bond name and functional group					
	carbonyl group C=O	Ester group stretching C–O	N–H amide protein in the polymer	C–H stretching methyl and methylene	–CH group	–OH group
			Wavelength cm^{-1}			
PHB	1720.862	—	—	2976.133	1278.590	—
+ve	—	1650.054	1536.889	2961.977	1275.334	3282.092
3/1	—	1650.212	1532.606	2961.507	1285.725	3283.034
9/1	—	—	1531.962	—	—	—
10/1	—	1649.884	1537.600	2163.595	1234.149	—
12/1	—	1632.607	1531.058	2922.235	1258.489	—
17/1	—	1633.190	1533.594	2961.555	1258.594	3282.994
19/1	—	1645.382	1537.159	2962.102	1285.469	—
20/1	—	1634.394	1530.941	2962.579	1285.454	3284.661
22/1	—	1643.354	1532.321	2961.308	1259.401	3280.077
25/1	—	1650.201	1533.100	2962.169	1285.953	3282.528
6/2	—	1650.268	1537.788	—	1233.206	—
8/2	—	1653.275	1532.055	2357.763	—	—
9/2	—	1637.733	1532.606	—	—	—
10/2	—	1646.277	1530.920	2922.692	1259.232	—
13/2	—	1650.393	1538.062	2357.890	1215.417	3272.019
16/2	—	1650.650	1536.901	2159.899	1236.901	—

Strain ID	Bond name and functional group					
	carbonyl group C=O	Ester group stretching C–O	N–H amide protein in the polymer	C–H stretching methyl and methylene	–CH group	–OH group
17/2	—	1638.042	1526.125	2965.548	1229.005	3297.117
18/2	—	1643.042	1537.981	2161.182	1234.679	3282.213
19/2	—	1635.190	1530.788	2361.236	1248.539	3277.935
24-1/2	—	1632.858	—	2933.581	1226.658	—
24/2	—	1635.579	—	2359.462	1249.115	3347.794

Table 3.
Functional groups identified by the FTIR method for PHA analysis.

The results of the FTIR analysis of the PHA extracted from the 20 environmental samples showed a resemblance to the positive strain's characteristic absorption peaks, which indicates the presence of the PHA functional groups in the samples. As it was mentioned above, the positive-control strain was confirmed as a PHA producer by PCR and the staining methods. The strain 9/1 was tested positive by Sudan Black B, negative for the *phaC* gene by the PCR method, and it did not show any characteristic absorption peaks corresponding to PHA production. As a result, it was concluded that this strain was not a PHA producer.

6. Biochip analysis of environmental samples for the presence of PHA-producing bacteria

A low-density biochip as a collection of DNA probes arranged on a solid matrix was the first time applied for rapid screening of PHA-producing organisms in the environment. This technique performs detection using a single undivided environmental sample and allows the identification of different genes in one reaction. The development of a biochip includes the following steps: design of the appropriate probes; equalization of the hybridization capacity of the probes for their inclusion in the biochip prototype; environmental DNA preparation for hybridization; and the fine-tuning of a biochip for the optimization conditions of bacteria detection.

Eleven oligonucleotide probes for the *phaC* genes responsible for the metabolic pathways involved in PHA production were designed (**Table 4**). The phaC gene codes

Species	Gene	Probe	Sequence 5′ → 3′	Probe length, bp
Burkholderia spp.	PHA synthase (phaC) gene poly-beta-hydroxybutyrate polymerase gene	Probe 1	TCA ACA AGT TCT ACA TCC TCG	21
Burkholderia spp.	poly-beta-hydroxybutyrate polymerase gene	Probe 2	CGT GCA TCA ACA AGT TCT ACA	21
Burkholderia pseudomallei	poly-beta-hydroxybutyrate polymerase gene	Probe 3	TCT GCG GAA TAC CTA TCT CGA	21

Species	Gene	Probe	Sequence 5′ → 3′	Probe length, bp
Synechococcus sp. MA19	PhaC (phaC) gene	Probe 4	TTT AGG TAA CAT TCG CAT GCC	21
Synechococcus sp. MA19	PhaC (phaC) gene	Probe 5	TTT AAT GCT CAA ACC CCG ACA	21
Pseudomonas sp.	Poly(R)-hydroxyalkanoic acid synthase	Probe 6	GCC ACA GAT CAA CAA GTT CTA	21
Pseudomonas sp.	Poly(R)-hydroxyalkanoic acid synthase	Probe 7	TGA TCT GGA ACT ACT GGG TCA	21
Pseudomonas chlororaphis	Poly(R)-hydroxyalkanoic acid synthase	Probe 8	CCG GGT ACT TAT GTC CAT GAA	21
Aquabacterium sp. A7-Y	PHA synthase (phaC) gene	Probe 9	TAT CTC GAA AAC AAG CTC AGC	21
Cupriavidus sp.	polyhydroxyalkanoate synthase (phaC) gene	Probe 10	ATG GCG ACC GGC AAG GGT GCG	21
Uncultured bacterium clone, Cupriavidus sp.	PHA synthase (phaC) gene poly-beta-hydroxybuterate polymerase	Probe 11	CGT GCA TCA ACA AGT ACT ACA	21

Table 4.
Probes used in the biochip.

a subunit common to the four known PHA synthase classes (class I, class II, class III, and class IV) [35], which are found in different species of bacteria; the PHA C polypeptide subunit varies in molecular weights. The BLAST search in the NCBI database (http://www.ncbi.nlm.nih.gov) was used to derive the sequences used for the probe design. The complete nucleotide sequence for the genomic DNA of PHA-producing microorganisms was downloaded and saved in FASTA format. The sequence was loaded into the SeaView software, which is a sequence alignment editor allowing manual or automatic alignment through an interface with the CLUSTALW program. Region variations in the sequences were identified from multiple alignments generated by the Clustalo option in SeaView [36] or creating multiple alignments of protein sequences by Muscle [37]. Based on final alignments, conserved parts of the sequences of each gene were chosen and used for probe designing using the OLIGO 7 program [38]. The phylogenetic tree generated after a distance-based analysis using the Kimura 2 model and neighbor-joining algorithm showed a clear distinction between the classes of *phaC* genes. After the phylogenetic analyses, the probes were designed to cover different classes of PHA synthase.

Probes (4–10) target PHA synthase (*phaC*) gene, probes (2,3) target poly-beta-hydroxybutyrate polymerase gene, and probes 1 and 11 target both genes.

The 3D dendrimeric matrixes for biochip preparation were manufactured at Tbilisi State University, Andronikashvili Institute of Physics, Georgia. All procedures concerning dendrimeric matrix activation for probes immobilization, and dendrimeric matrix deactivation for environmental DNA hybridization are presented in our previous publication [39].

In order to check and evaluate that the chosen probes have similar hybridization capacities, the cassette approach developed and published in Ref. [40] has been used. If the probes exhibit the same hybridization characteristics, they should give equal fluorescent signals in hybridization reactions. The result of the estimation of the hybridization capacity of the proposed probes showed that the probes exhibited comparable hybridization signals and were characterized by the mean signal-to-noise ratio S/N = 12 (data not shown). These probes were tested against DNA obtained from environmental samples (sludge and compost) samples. Total DNA was purified from the environmental samples (sludge or compost) using MO BIO PowerSoil® DNA Isolation Kit (MO BIO technologist, Inc.). The procedures for DNA preparation for hybridization that include DNA amplification, fragmentation, and fluorescent labeling are described in Ref. [40]. Two sludge and three compost environmental samples were analyzed for the bacterial assemblage, revealing PHAs producing capacity using a biochip method. **Figure 7** represents the proportion/interrelation of the issued by the probes´ bacterial species in the functional assemblage of sludge T2, as the equalized biochip is the basis for the estimation of the microbial ratio. The hybridization signal S/N above 1.5 was counted as a reliable result.

According to **Figure 7** and **Table 4**,

- Probes 1, 2, and 3 are probes for the identification of *Burkholderia* sp. For all of them, the identification on the biochip is under the threshold of detection (S/N < 1.5).

- Probes 4 and 5 are different probes for the identification of *Synechococcus* sp. MA19. The probes are characterized by identical hybridization capacity (cassette assay); however, probe 5 is more efficient in the case of real ecological samples.

- Probes 6, 7, and 8 are probes for the identification of *Pseudomonas* sp. All of them are characterized by S/N > 1.5.

- Probe 9 is a probe for the identification of *Aquabacterium* sp. A7-Y. The hybridization signal is reliable (S/N > 1.5).

- Probes 10 and 11 are probes for the identification of *Cupriavidus* sp. The identification of this bacteria on the biochip by probe 10 is reliable, but by probe 11 under the threshold of detection (S/N < 1.5).

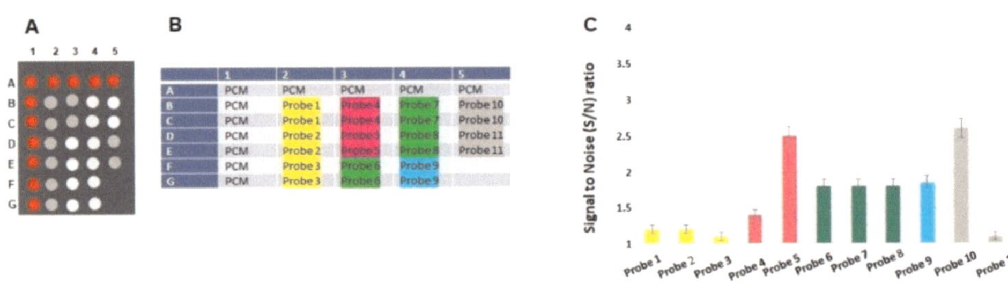

Figure 7.
The results of DNA from T2 sludge sample hybridization with the biochip. Panels A and B represent the arrangement of the probes on the biochips. The red circles and PCM are the position control markers. The white circles mark the position of the reliably visible signals detected after hybridization with the probes for phaC gene. Panel C shows the hybridization signal, presenting a signal-to-noise ratio, for the probes on the biochip. The data presented are mean values ± SD.

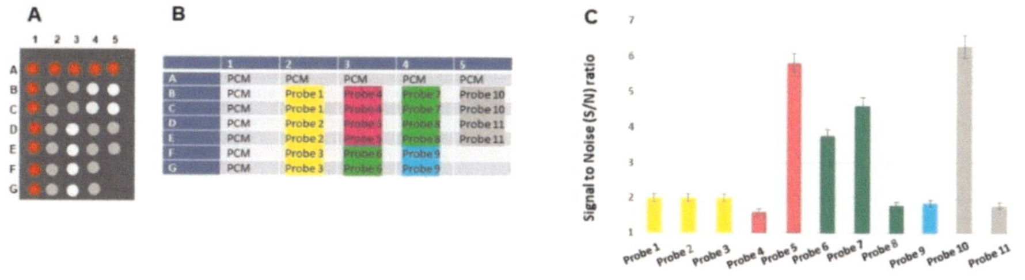

Figure 8.
The results of DNA from T5 compost sample hybridization with the biochip. Panels A and B represent the arrangement of the probes on the biochips. The red circles and PCM are the position control markers. The white circles mark the position of the reliably visible signals detected after hybridization with the probes for phaC gene. Panel C shows the hybridization signal, presenting as a signal-to-noise ratio, for the probes on the biochip. The data presented are mean values ± SD.

Therefore, the bacteria species present in sample T2 are *Synechococcus*; *Pseudomonas*; *Cupravidus,* and *Aquabacterium*. The first three species are very well-known PHA producers.

Figure 8 represents the proportion of the bacterial species in the functional assemblage of compost T5.

According to **Figure 8** and **Table 4**,

- Probes 1, 2, and 3 are probes for the identification of *Burkholderia* sp. For all of them, the identification on the biochip is above the threshold of detection (S/N > 1.5).

- Probes 4 and 5 are different probes for the identification of *Synechococcus* sp. MA19. The probes are characterized by identical hybridization capacity (cassette assay); however, probe 5 is again more efficient in the case of real ecological samples.

- Probes 6, 7, and 8 are probes for the identification of *Pseudomonas* sp. All of them are characterized by S/N > 1.5; however, the signals from probes 6 and 7 are much stronger than for probe 8.

- Probe 9 is a probe for the identification of *Aquabacterium* sp. A7-Y. The hybridization signal is reliable (S/N > 1.5).

- Probes 10 and 11 are probes for the identification of *Cupriavidus* sp. The signal from both probes is reliable; however, the signal of probe 10 on the biochip is again much stronger than of probe 11.

Therefore, all studied types of bacteria are present in sample T5. Dominant bacteria species in sample T5 are *Synechococcus*, *Pseudomonas*, and *Cupravidus*, well-known PHA producers.

Table 5 summarizes the result obtained with biochips for the studied environmental samples.

The compost samples T5 and T6 contain all studied bacteria species with *phaC* gene. The bacteria spp. *Synechococcus* and *Cupravidus* (probes number 5 and 10) were detected in all environmental samples (sludge and composts). The absence of hybridization signals on the biochip for some bacteria might be explained by the absence of these bacteria in the samples.

	T2 sludge	T4 sludge	T5 compost (food-based waste)	T6 compost (warm garden waste)	T7 compost (graveyard)
Probe 1	−	+	+	+	−
Probe 2	−	−	+	+	−
Probe 3	−	−	+	−	−
Probe 4	−	−	+	+	+
Probe 5	+	+	+	+	+
Probe 6	+	+	+	+	−
Probe 7	+	−	+	+	+
Probe 8	−	−	+	+	−
Probe 9	+	−	+	+	−
Probe 10	+	+	+	+	+
Probe 11	−	−	+	+	−

Table 5.
Summary of the phaC gene detected in the biochip from environmental samples.

7. Conclusions

Bioplastics are dramatically favored over oil-based plastics due to their ability to degrade. The use of such natural polymers would help counteract the current accumulation of standard nonbiodegradable polymers in the global environment. Bioplastics have many applications, such as packaging materials, biomedical implants, drug delivery systems, and biofuels. The screening of the indigenous bacteria in the environmental samples is the first key step on the way of the selection of bioplastic-producing organisms. In this study, the conventional chemical, molecular biological, and physical methods were successfully applied for the screening of the environmental bacteria-producing PHAs, biopolymers that can be used to produce bioplastics. Twenty bacterial strains have tested positive for *phaC* genes using the PCR method and/or the Sudan Black B, Nile Blue, and Nile Red staining methods. The PHAs were extracted from these bacterial strains and characterized using the FTIR method. The main FTIR spectral peaks for the PHAs extracted from 19 studied bacterial isolates resemble the peaks for the PHAs isolated from PHA-producing bacterial strain (+ve) that was used as a positive-control strain, indicating the presence of the PHA functional groups in the samples. Diagnostic biochip was first time explored as a fast method for primary screening of PHA-producing microorganisms in environmental samples. The hybridization results showed that the designed probes successfully detected the *phaC* genes, which covered three classes of PHA synthase in the environmental samples. Class I resides in *Cupriavidus* sp.; classes I and II are detected in *Pseudomonas* sp., and class III are found in the *Synechococcus* and *Burkholderia* spp. The bacterial composition in the environmental samples revealed the frequently encountered *Synechococcus*, *Pseudomonas*, and *Cupravidus* species, well-known PHA producers. Moreover, the use of biochips and staining procedures for the detection of biopolymers-producing bacteria will allow the screening of microorganisms for novel PHA production pathways.

Author details

Aisha S. Alwuhaib[1], Vitaly Zinkevich[2,3], Tamar Kartvelishvili[3], Nino Asatiani[3] and Nelly Sapojnikova[3*]

1 Research and Development Center, Saudi Aramco, Dhahran, Saudi Arabia

2 School of Pharmacy and Biomedical Sciences, University of Portsmouth, UK

3 I. Javakhishvili Tbilisi State University, Andronikashvili Institute of Physics, Tbilisi, Georgia

*Address all correspondence to: nelly.sapojnikova@tsu.ge

References

[1] Malla-Pradhan R, Pradhan BL, Phoungthong K, Joshi TP. Occurrence and distribution of microplastics from Nepal's second largest lake. Water, Air, and Soil Pollution. 2022;**233**:423. DOI: 10.1007/s11270-022-05896-z

[2] Anjum A, Zuber M, Zia K, Noreen A, Anjum M, Tabasum S. Microbial production of polyhydroxyalkanoates (PHAs) and its copolymers: A review of recent advancements. International Journal of Biological Macromolecules. 2016;**89**:161-174. DOI: 10.1016/j.ijbiomac.2016.04.069

[3] Grage K, Peters V, Palanisamy R, Rehm BHA. Polyhydroxyalkanoates: From bacterial storage compound via alternative plastic to bio-bead. In: Rehm BHA, editor. Microbial Production of Biopolymers and Polymer Precursors: Applications and Perspectives. Norfolk: Caister Academic Press; 2009

[4] Adkins J, Pugh S, McKenna R, Nielsen DR. Engineering microbial chemical factories to produce renewable "biomonomers". Frontiers in Microbiology. 2012;**3**:313. DOI: 10.3389/fmicb.2012.00313

[5] Danial AW, Hamdy SM, Alrumman SA, Gad El-Rab SMF, Shoreit AAM, Hesham AE-L. Bioplastic production by *Bacillus wiedmannii* AS-02 OK576278 using different agricultural wastes. Microorganisms. 2021;**9**:2395. DOI: 10.3390/microorganisms9112395

[6] Matsumoto K, Murata T, Nagao R, Nomura C, Arai S, Arai Y, et al. Production of short-chain-length/medium-chain-length Polyhydroxyalkanoate (PHA) copolymer in the plastid of *Arabidopsis thaliana* using an engineered 3-ketoacyl-acyl carrier protein synthase III. Biomacromolecules. 2009;**10**:686-690. DOI: 10.1021/bm8013878

[7] Singh A, Mallick N. SCL-LCL-PHA copolymer production by a local isolate, *Pseudomonas aeruginosa* MTCC 7925. Biotechnology Journal. 2009;**4**:703-711. DOI: 10.1002/biot.200800307

[8] Bugnicourt E, Cinelli P, Lazzeri A, Alvarez V. Polyhydroxyalkanoate (PHA): Review of synthesis, characteristics, processing and potential applications in packaging. Express Polymer Letters. 2014;**8**:791-808. DOI: 10.3144/expresspolymlett.2014.82

[9] Madison LL, Huisman GW. Metabolic engineering of poly(3-Hydroxyalkanoates): From DNA to plastic. Microbiology and Molecular Biology Reviews. 1999;**63**:21-53. DOI: 10.1128/MMBR.63.1.21-53.1999

[10] Sehgal R, Gupta R. Polyhydroxyalkanoate and its efficient production: An eco-friendly approach towards development. 3 Biotech. 2020;**10**:549-563. DOI: 10.1007/s13205-020-02550-5

[11] Lau NS, Sudesh K. Revelation of the ability of *Burkholderia* sp. USM (JCM 15050) PHA synthase to polymerize 4-hydroxybutyrate monomer. AMB Express. 2012;**2**:41. DOI: 10.1186/2191-0855-2-41

[12] Kanavaki I, Drakonaki A, Geladas ED, Spyros A, Xie H, Tsiotis G. Polyhydroxyalkanoate (PHA) production in *Pseudomonas* sp. phDV1 strain grown on phenol as carbon sources. Microorganisms. 2021;**9**:1636. DOI: 10.3390/microorganisms9081636

[13] Chen G-Q. Plastics completely synthesized by Bacteria:

Polyhydroxyalkanoates. In: Chen G-Q, editor. Plastics from Bacteria: Natural Functions and Applications. Berlin, Heidelberg: Springer; 2010. pp. 1-17

[14] Brigham C, Speth D, Rha C, Sinskey A. Whole-genome microarray and gene deletion studies reveal regulation of the Polyhydroxyalkanoate production cycle by the stringent response in *Ralstonia eutropha* H16. Applied and Environmental Microbiology. 2012;**78**:8033-8044. DOI: 10.1128/AEM.01693-12

[15] Koller M, Bona R, Braunegg G, Hermann C, Horvat P, Kroutil M, et al. Production of Polyhydroxyalkanoates from agricultural waste and surplus materials. Biomacromolecules. 2005;**6**: 561-565. DOI: 10.1021/bm049478b

[16] Chaudhry WN, Jamil N, Ali I, Ayaz MH, Hasnain S. Screening for polyhydroxyalkanoate (PHA)-producing bacterial strains and comparison of PHA production from various inexpensive carbon sources. Annals of Microbiology. 2010;**61**:623-629. DOI: 10.1007/s13213-010-0181-6

[17] Pagliano G, Galletti P, Samorì C, Zaghini A, Torri C. Recovery of Polyhydroxyalkanoates from single and mixed microbial cultures: A review. Frontiers in Bioengineering and Biotechnology. 2021;**9**:624021. DOI: 10.3389/fbioe.2021.624021

[18] Spiekermann P, Rehm B, Kalscheuer R, Baumeister D, Steinbüchel A. A sensitive, viable-colony staining method using Nile red for direct screening of bacteria that accumulate polyhydroxyalkanoic acids and other lipid storage compounds. Archives of Microbiology. 1999;**171**:73-80. DOI: 10.1007/s002030050681

[19] Ostle AG, Holt JG. Nile blue A as a fluorescent stain for poly-beta-hydroxybutyrate. Applied and Environmental Microbiology. 1982;**44**: 238-241. DOI: 10.1128/aem.44.1.238-241.1982

[20] Steinbuchel A, Frund C, Jendrossek D, Schlegel HG. Isolation of mutants of *Alcaligenes eutrophus* to derepress the fermentative ealcohol-dehydrogenase. Archives of Microbiology. 1987;**148**:178-186. DOI: 10.1007/BF00414809

[21] Legat A, Gruber C, Zangger K, Wanner G, Stan-Lotter H. Identification of polyhydroxyalkanoates in *Halococcus* and other haloarchaeal species. Applied Microbiology and Biotechnology. 2010; **87**:1119-1127. DOI: 10.1007/s00253-010-2611-6

[22] Mahansaria R, Choudhury JD, Mukherjee J. Polymerase chain reaction-based screening method applicable universally to environmental haloarchaea and halobacteria for identifying polyhydroxyalkanoate producers among them. Extremophiles. 2015;**19**:1041-1054. DOI: 10.1007/s00792-015-0775-9

[23] Martinez V, Henary M. Nile red and Nile blue: Applications and syntheses of structural analogues. Chemistry—A European Journal. 2016;**22**:13764-13782. DOI: 10.1002/chem.201601570

[24] Solaiman D, Ashby R, Foglia T. Rapid and specific identification of medium-chain-length polyhydroxyalkanoate synthase gene by polymerase chain reaction. Applied Microbiology and Biotechnology. 2000; **53**:690-694. DOI: 10.1007/s002530000332

[25] Huang Y-T, Chen P-L, Semblante GU, You S-J. Detection of

Polyhydroxyalkanoate-accumulating Bacteria from domestic wastewater treatment plant using highly sensitive PCR primers. Journal of Microbiology and Biotechnology. 2012;**22**:1141-1147. DOI: 10.4014/jmb.1111.11040

[26] Kung S, Chuang Y, Chen C, Chien C. Isolation of polyhydroxyalkanoates-producing bacteria using a combination of phenotypic and genotypic approach. Letters in Applied Microbiology. 2007; **44**:364-371. DOI: 10.1111/j.1472-765X.2006.02090.x

[27] Sheu D-S, Wang Y-T, Lee C-Y. Rapid detection of polyhydroxyalkanoate-accumulating bacteria isolated from the environment by colony PCR. Microbiology. 2000;**146**:2019-2025. DOI: 10.1099/00221287-146-8-2019

[28] Turner S, Pryer KM, Miao VPW, Palmer JD. Investigating deep phylogenetic relationships among cyanobacteria and plastids by small subunit rRNA sequence analysis. The Journal of Eukaryotic Microbiology. 1999;**46**:327-338. DOI: 10.1111/j.1550-7408.1999.tb04612.x

[29] Lane D, Weisburg W, Barns S, Pelletier D. 16S ribosomal DNA amplification for phylogenetic study. Journal of Bacteriology. 1991;**173**: 697-703. DOI: 10.1128/jb.173.2.697-703.1991

[30] Arshad A, Ashraf B, Ali I, Jamil N. Biosynthesis of polyhydroxyalkanoates from styrene by *Enterobacter* spp. isolated from polluted environment. Frontiers in Biology. 2017;**12**:210-218. DOI: 10.1007/s11515-017-1446-2

[31] Apparao U, Krishnaswamy VG. Production of Polyhydroxyalkanoate (PHA) by a moderately halotolerant bacterium *Klebsiella pneumoniae* U1 isolated from rubber plantation area. International Journal of Environmental Bioremediation and Biodegradation. 2015;**3**:54-61. DOI: 10.12691/ijebb-3-2-3

[32] Isak I, Patel M, Riddell M, West M, Bowers T, Wijeyekoon S, et al. Quantification of polyhydroxyalkanoates in mixed and pure cultures biomass by Fourier transform infrared spectroscopy: Comparison of different approaches. Letters in Applied Microbiology. 2016; **63**:139-146. DOI: 10.1111/lam.12605

[33] Sindhu R, Ammu B, Binod P, Deepthi S, Ramachandran K, Soccol C, et al. Production and characterization of poly-3-hydroxybutyrate from crude glycerol by *Bacillus sphaericus* NII 0838 and improving its thermal properties by blending with other polymers. Brazilian Archives of Biology and Technology. 2011;**54**:783-794. DOI: 10.1590/S1516-89132011000400019

[34] Tufail S, Munir S, Jamil N. Variation analysis of bacterial polyhydroxyalkanoates production using saturated and unsaturated hydrocarbons. Brazilian Journal of Microbiology. 2017; **48**:629-636. DOI: 10.1016/j.bjm.2017.02.008

[35] McCool G, Cannon M. PhaC and PhaR are required for Polyhydroxyalkanoic acid synthase activity in *Bacillus megaterium*. Journal of Bacteriology. 2001;**183**:4235-4243. DOI: 10.1128/JB.183.14.4235-4243.2001

[36] Sievers F, Wilm A, Dineen D, Gibson T, Karplus K, Li W, et al. Fast, scalable generation of high-quality protein multiple sequence alignments using Clustal omega. Molecular Systems Biology. 2014;7:539-539. DOI: 10.1038/msb.2011.75

[37] Edgar R. MUSCLE: Multiple sequence alignment with high accuracy

and high throughput. Nucleic Acids Research. 2004;**32**:1792-1797. DOI: 10.1093/nar/gkh340

[38] Rychlik W. OLIGO 7 primer analysis software. In: Yuryev A, editor. PCR Primer Design. Methods in Molecular Biology. Totowa: Humana Press; 2007. pp. 35-59

[39] Asatiani N, Abuladze M, Kartvelishvili T, Osepashvili M, Shengelaya A, Daraselia D, et al. Copper (II) ion action on soil Bacteria. Water, Air, and Soil Pollution. 2021;**232**:355. DOI: 10.1007/s11270-021-05317-7

[40] Zinkevich V, Sapojnikova N, Mitchell J, Kartvelishvili T, Asatiani N, Alkhalil S, et al. A novel cassette method for probe evaluation in the designed biochips. PLoS One. 2014;**9**:e98596. DOI: 10.1371/journal.pone.0098596

Chapter 3

Preparation and Analysis of Standard Microplastics

Raffaella Mossotti, Giulia Dalla Fontana, Anastasia Anceschi, Enrico Gasparin and Tiziano Battistini

Abstract

Over 14 million tons of microplastic have been accumulated in water resources and they are increasing yearly. About 8% of European microplastic released into the water are from synthetic textiles. This kind of microplastic is generally in the form of microfilaments. They have a higher potential to enter the food chain due to their size and shape. Although microfilaments generate great concern, no precise guidelines for their quantification and qualification are yet available. Thus, in this chapter, the origin of microfilaments is fully investigated. After that, a novel approach for identifying and counting microplastic with fiber shape is presented. An accurate method for preparing microfilament standard suspensions is described to facilitate lab tests and have a reliable methodology for monitoring microplastic pollution.

Keywords: textiles, synthetic thread, microplastics with fiber shape, standard suspension, quantification

1. Introduction

This chapter presents a reliable method for preparing standard microfilaments to facilitate lab testing and monitoring of microplastics in different matrices. The scope is to achieve a positive impact on the quality control of all operations, from sampling to counting and identification. Using standard synthetic microfilaments as references for the validation of common experimental procedures could reduce differences between data. Furthermore, a standard synthetic fiber material would allow the monitoring of the ecotoxicological impacts of microplastics on biota and human health in line with the European Commission's Green Deal, the Circular Economy Action Plan and the proposed remedial actions supported by the U.N SDGs under Goal 14.

2. What is plastic?

Plastics are defined as synthetic organic polymers typically made from petrochemicals.

Specifically, synthetic polymer molecules consist of many monomers which react in different ways. Many simple hydrocarbons, such as ethylene and propylene, can be transformed into polymers by adding monomers to the growing chain. The combination of these monomers creates various kinds of polymers. Other substances can be added to polymers to give the final product some desired characteristics [1].

Sometimes, the term plastic is also used to indicate blends with different synthetic polymers or other low molecular weight compounds such as additives, UV or thermal stabilizers, flame retardants, dyes, antioxidants, plasticizers, etc. [2].

Because plastics are considered chemically, physically, biologically stable and resistant materials, once in the environment, they can undergo degradation upon exposure to different factors, such as sunlight, water, and wind, and break down into tiny plastic particles known as microplastics. After fragmentation, they can be transported by wind and water due to their lightweight [3].

Thus, once in the environment, microplastics accumulate and persist. Consequently, they are ubiquitous in terrestrial, fresh water and marine environments [4]. The source of microplastics includes wastewater treatment plants, landfills, automotive tires, pre-production plastic pellets, synthetic clothing, road signs, and paint [5], (**Figure 1**).

Among all sectors, the textile one is considered one of the major sources of microplastic pollution.

Textile processes are responsible for 20% of global water pollution and the washing of synthetic garments contributes to about 35% of the global release of primary microplastics. These materials are not retained during the filtration systems of wastewater treatment plants (WWTPs) and, therefore, enter the marine ecosystem directly [6].

These microplastics occur in different forms (e.g., cylindrical, spherical, etc.) and partly escape the filtration systems of WWTPs, reaching seas and oceans directly. In this respect, the identification and classification of fiber fragments are necessary to spot any weak points in the textile production process and in the life cycle of synthetic garments. The release of microplastics can occur during the different processes and use phases, including spinning, weaving, finishing (gauzing, finishing, dyeing), packaging, wear, washing, drying, and finally, at the end of life, landfill disposal [7].

Figure 1.
Example of microplastic sources in water system (https://unsplash.com).

Figure 2.
Plastics spread in the environment (https://unsplash.com).

Hartline et al. have estimated that a WWTP plant with a 94% of removal percentage and considering 0.35 m^3 of wastewater per person a day would produce for 100,000 people about 1.02 kg of microfilaments (**Figure 2**) [8].

The term microplastics was first coined by R. Thomson in 2004 by observing micrometer-sized plastic fragments in marine sediments and then refined by Arthur et al. by setting size limits above 5 mm [9, 10]. Later in 2011, Cole et al. divided microplastics into two categories: primary ones produced at a micro size and secondary ones that only reach that size through fragmentation and degradation due to environmental biodegradation effects. In 2016, nano-sized particles were also included in the definition of microplastics GESAMP [11].

Although the definition of microplastics is still being debated, the current one follows the European Chemical Agency (ECHA), "a material composed of solid polymeric-containing particles to which additives or other substances may be added. The family of microplastics includes synthetic-based particles, such as polyethylene (PE), polypropylene (PP), polystyrene (PS), polyamides (PA), polyethylene terephthalate (PET), polyvinyl chloride (PVC), polyacrylonitrile (PAN), polymethylacrylate (PMA), elastomers and silicone rubber with particles ranging from 1 nm to 5 mm and

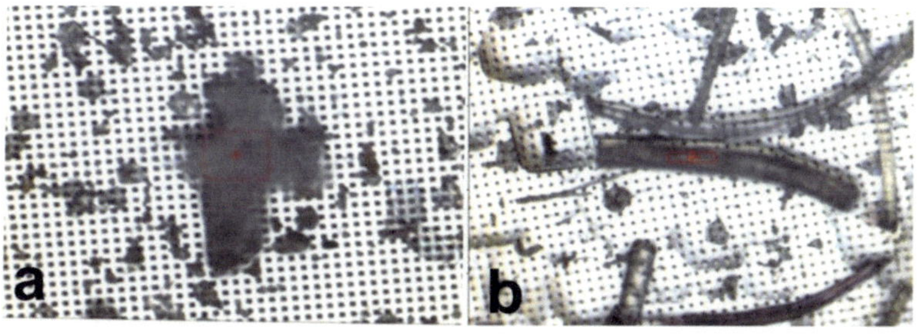

Figure 3.
Optical images obtained with an optical microscope coupled to MicroFTIR of a) polypropylene microparticle and b) polyester microfilaments.

fiber lengths ranging from 3 nm to 15 mm and a length-to-diameter ratio of > 3" [12]. The difference between particles and filaments is reported in **Figure 3**.

2.1 Textile fibers

Fibers are a class of materials consisting of a fibrous structure whose length is thousands of times higher than its diameter. Fibers are the units from which all textile materials are made. They are incredibly important to textile production as they have properties that allow them to be spun into yarn or directly made into fabric. This means they must be strong enough to hold their shape, flexible enough to be shaped into a fabric or yarn, elastic enough to stretch, and durable enough to last. Textile fibers also have to be a minimum of 5 millimeters in length as shorter ones cannot be spun together. Textile fibers are generally classified as natural or man-made. An outline is reported in **Figure 4**.

Natural fibers are further subdivided into animal (e.g., wool, mohair, cashmere, angora, silk), vegetable (e.g., cotton, flax, kapok, jute, hemp), or mineral (e.g., asbestos), as shown in **Figure 4**. Animal fibers are typically obtained from the coats or fleeces of animals, or in the case of silk, the raw material is the extruded filaments of the silkworm cocoon [13]. Vegetable fibers grow as seeds, leaves and bast fibers, whereas mineral ones are mainly asbestos fibers. In **Figure 5** an optical picture of animal (a) and vegetable (b) fibers is reported.

Wool fibers have the form of elliptical cylinders. The range diameter of around 20 μm is typical of merino wool, the most commonly used for clothes. It shows a scale

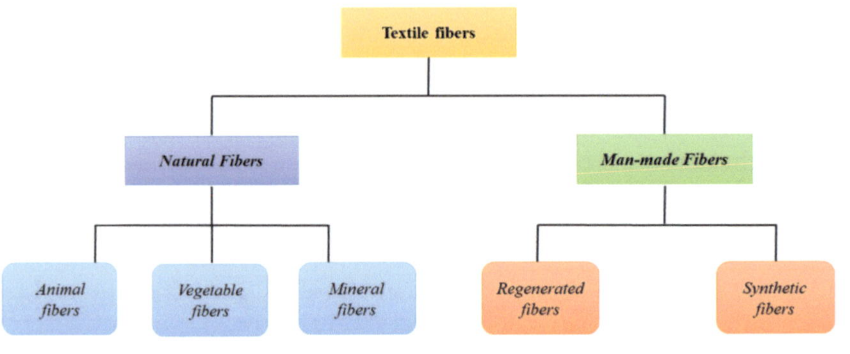

Figure 4.
Classification of textile fibers.

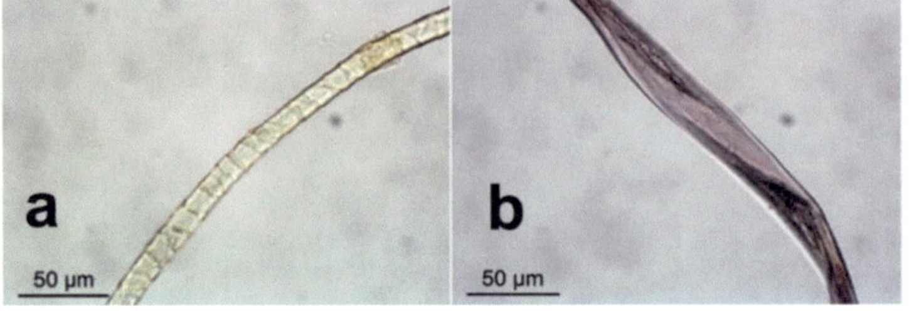

Figure 5.
Optical microscopy image of merino loose wool (a) and cotton fibers (b) obtained in transmission mode. (500 X).

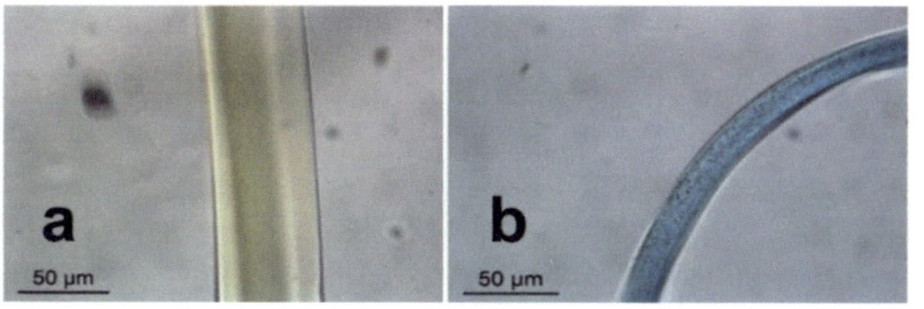

Figure 6.
Optical microscopy image of PA 6 (a) and PA 6.6 (b) fibers obtained in transmission mode (500 X).

structure with an irregular profile and a stopped one due to overlapped cuticle cells, like a tiled roof. Cotton fibers show a flat band structure with corkscrew-like twisting. The convolution frequently varies between 3.9 and 6.5 per mm and the number of reversals per mm ranges between 1.0 and 1.7. The longest cotton fiber is 2.8 cm and can be found in Scottish thread. The section of the fibers shows variable shapes such as elliptical, oval, and kidney with a well-defined central lumen parallel to the outer wall [14].

Man-made fibers are any fiber that is derived from an artificial process. The fibers made from chemical synthesis are called synthetic fibers, e.g., Polyamide 6 (PA 6), Polyamide 6.6 (PA 6.6), polyester (PET), polyacrylic (PAN), cholorofibers (PVC), and aramids (kevlar), while the fibers generated from natural polymer sources are called regenerated fibers or natural polymer fibers e.g.: Viscose, Rayon, cellulose triacetate, etc. [13]. An optical micrograph of synthetic fibers is reported in **Figure 6**.

The fibers are uniaxially oriented during the melt, dry, or wet spinning process, which gives the fibers high tenacity and strength. Typically, they appear as smooth filaments, cylindrical or slightly elliptical.

Figure 6 and b shows the morphology of two nylon fibers (PA6 versus PA 6.6). [15]. They are generally semicrystalline polymers extruded and drawn in various cross-sectional shapes, which can be circular, kidney-shaped or three-lobed with smooth edges. In **Figure 6b**, the fiber shows the presence of fillers.

2.2 Why do fibers from clothes pollute?

Man-made fibers have tripled their market share from 23% in 1965 to nearly 72%. In addition, synthetic fibers have continued to grow to 75%, while cellulosic fibers, for example, have remained constant at about 6.4% [15]. Compared to natural ones, synthetic fibers do not depend on animal breeding or cultivation and are not affected by environmental factors such as seasonality and climate change.

Polyester is considered the best fiber in terms of production cost, raw material quality, and ability to improve performance and properties. Polyester fibers have reached 85% of the market share of the synthetic sector [16].

Moreover, in recent years, synthetic fibers have become the main protagonists of fast fashion (a clothing industry that produces low-quality and low-priced clothing and constantly launches new collections in a short time), generating large amounts of waste from unsold, unwanted and/or landfilled goods.

Furthermore, synthetic textiles are estimated to be responsible for a global discharge of between 0.2 and 0.5 million tons of microplastics into the oceans yearly [17]. Synthetic fragments can enter the aquatic environment during use, machine washing and drying of garments, or through leaching of waste material (pre-consumer, post-consumer) that accumulates in landfills.

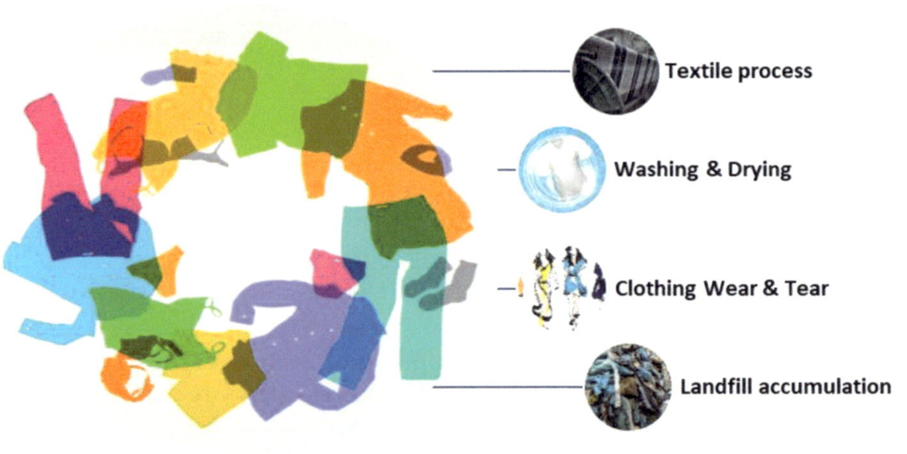

Figure 7.
Source of microplastic fibers release during textile life-cycle.

According to [18] approximately 35% of microplastics released into oceans globally originate from washing synthetic textiles, as shown by their incidence in freshwater and saline environments, near urban centers, in sewage sludge and its by-products, in wastewater treatment plant effluents, in sediments and in some biota such as invertebrates, birds, and fish.

Although wastewater from washing machines is considered the main transport route for synthetic microfilaments, air can be a possible way, too. The fibrous fragments are comparable to dispersed solid particles suspended in the air. Several researchers have pointed out that textile products, especially during manufacture, packaging, drying and use, can release microplastics. Furthermore, synthetic textiles used for upholstered furniture can release fibrous microplastics through friction and abrasion. Many works have shown that the amounts released are comparable to those produced during a washing machine cycle [19–21], as shown in **Figure** 7.

In recent years, concerns have grown about the environmental and health impacts associated with microplastic pollution. Textiles made of fibers of natural origin shed micro fragments, too. All fibers undergo a biodegradation process in water. However, natural fibers (e.g., cotton) are completely degraded in the aqueous matrix, whereas in the case of synthetic fibers, there is no complete degradation but fragmentation

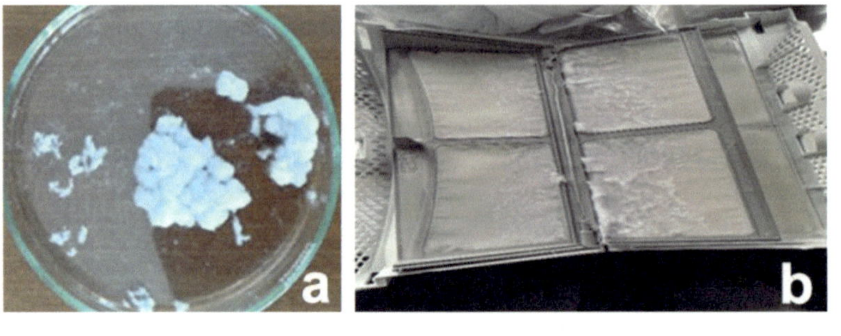

Figure 8.
Example of fiber material released from: a) the synthetic fabric during a 40-minute washing cycle (Wash & Wear) in laundry machines; b) tumble dryers (60-minute drying time).

into smaller filaments or particles that can reach nanometric dimensions. Another finding from the experimental data is that PET fibers are the most commonly found in the environment, followed by PAN, PP, and PA fibers (**Figure 8**) [22, 23].

2.3 Environmental impact

In recent years, a growing concern about microplastic environmental pollution and health impacts has emerged. Several studies have shown a certain degree of chronic exposure to microplastic pollution is an integral part of contemporary life [24]. Due to their shape, microplastic can be ingested by all living organisms, from plankton to humans. Furthermore, another source of concern is the potentially toxic chemicals that they can contain, such as additives, monomers, catalysts and other by-products. Once microplastics have been released into the environment, due to their fragmentation, degradation and chemical contents, they can reach the biota and consequently enter the food chain. In addition, microplastics have characteristics such as size, shape, polymer composition and even color that can potentially be more important than their concentration in the environment to induce adverse effects, making it more challenging to identify their impact on organisms. In addition, fibrous microplastic fragments in terms of size (length and diameter) geometry, physical properties and surface characteristics may be responsible for the levels of biological interfaces with tissues and cause pathology. Small microplastics can easily penetrate cells and organs and carry a considerable content of harmful substances due to the high surface area unit they possess [24, 25].

2.4 Microplastic textiles: related problems

Nowadays, estimating and measuring the quantities of microplastics, particularly those with fiber shapes, is challenging. Estimating the number of released microplastics is highly uncertain because of the lack of standardized sampling and measurement methods. Furthermore, the obtained data are not fully shared by the scientific community and are not validated with inter-laboratory tests. At present, the experimental and the analytical protocols under study are mainly focused on the determination of microplastic with particle shapes, leaving out fiber-shaped ones.

Indeed, microplastic textile standard methods are rarely used in the study cases. Existing methods for preparing MFs (microfilaments) are focused on cutting or cryogenically grinding synthetic filaments, resulting in a wide distribution of fiber lengths [26]. Some scientists have prepared nylon, polyethylene terephthalate (PET) and polypropylene (PP) microplastic fibers with pre-determined lengths (40, 70 or 100 μm) using a cryotome protocol. They proved that this method effectively produces tens of thousands of MFs suitable for testing [27].

Despite these promising results, the proposed analytical techniques have several drawbacks since they are limited in counting and separation.

Thus, a novel approach to counting and identifying fibrous microplastics is becoming fundamental. For this reason, a standardized analytical method and its subsequent validation must be obtained.

A possible solution to this lack could be the use of appropriate standard microfilaments. The more specific issues are microplastic cut-off size, sample type, sampling procedure, laboratory sample processing, identification techniques and reporting units. Therefore, a new routine for qualitative and quantitative microplastic analysis with fiber shape could be established to have a standardized analytical method to compare different lab results.

3. Strategy to solve the problem

Since microfilaments of standard material are not commercially available, a possible solution for the determination of microplastics could be the preparation and analysis of standard microfilaments in aqueous suspensions. This reliable method can help laboratories to monitor the quality of their analytical procedures. The advantage of such a procedure is that it is possible to produce different types of microfilaments with a narrow size distribution as well as blend them. This protocol could fill the gap in the knowledge of the identification and quantification of fibrous microplastics in textile or environmental matrices. In particular, the proposed procedure achieves the following objectives:

- The preparation of suspensions of known concentration of standard synthetic microfilaments, representative of the textile industry.
- The use of microfilament suspensions as an internal standard during the analysis of a real sample to monitor the quality of all operations and analyses.
- The preparation of suspensions of known concentration that can be used for inter-laboratory and inter-calibration tests.

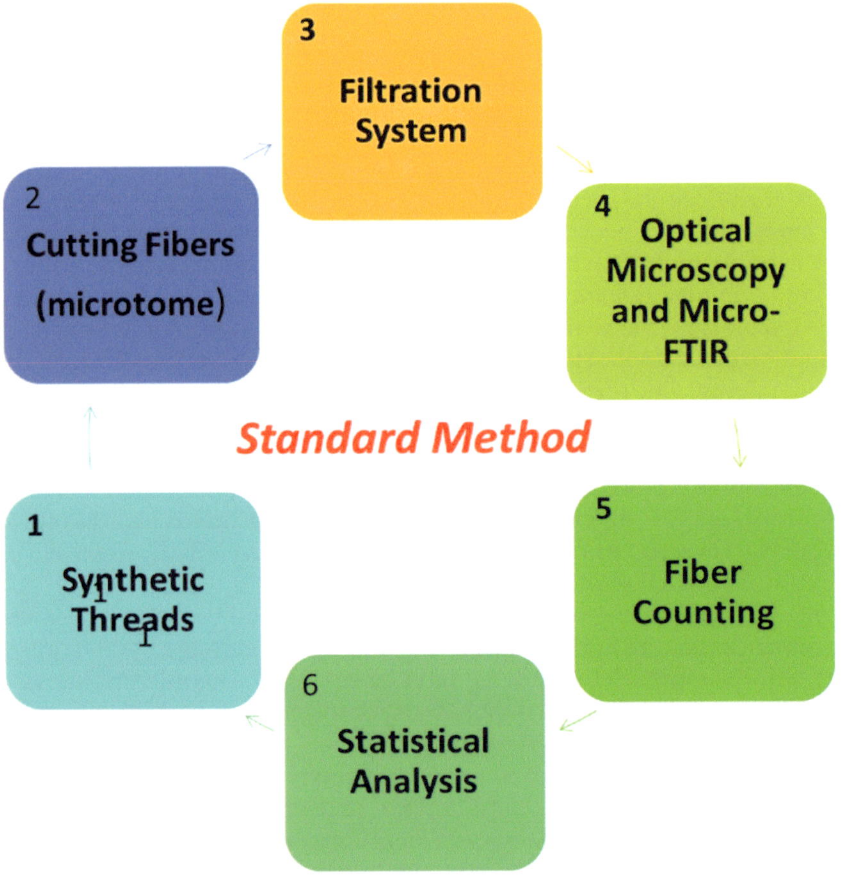

Figure 9.
Schematic diagram of the standard method steps.

- Identification, counting and analysis of fibrous microplastics in aqueous and non-textile aqueous matrices (**Figure 9**).

Mossotti et al. [28] developed a user-friendly method to prepare microfilament standard suspensions that can facilitate lab tests. Specifically, different synthetic threads of PA 6, PA 6.6, PET, and PP, which are shown in **Figure 10**, were used for the preparation of standard suspensions. They are commercial materials supplied by Aquafil S.p.A with a known number of filaments.

The parameters associated with all the yarns are: 1) Yellow PA 6 (180 filaments; 3450 dtex). 2) Blue PA 6.6 (68 filaments; 200 dtex). 3) Cream PET (256 filaments; 2970 dtex). 4) Orange PP (72 filaments; 70 dtex). An example of synthetic thread is shown in **Figure 11**.

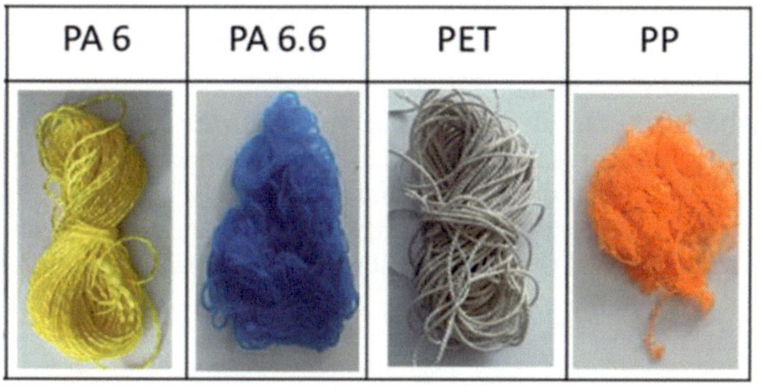

Figure 10.
Image of synthetic threads used for the preparation of the standard solution.

Figure 11.
a) Yarn; b) filaments; c) single filaments.

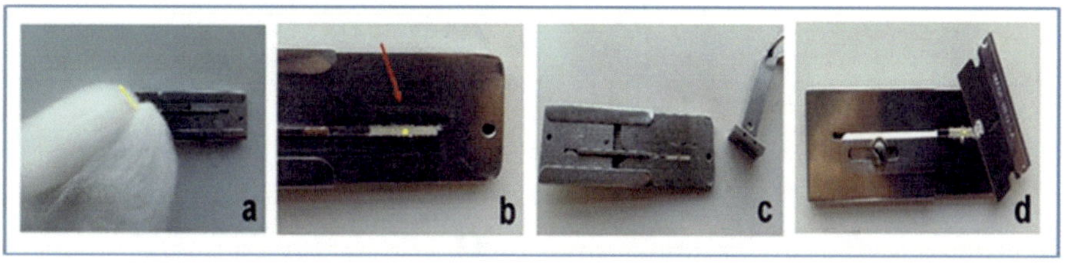

Figure 12.
a) Standard fibers and wool placed in a microtome slide; b) the protruding fringe removed by razor blade b) the fiber length chosen using a suitable pusher d) the cut fibers measure about 200 μm.

All synthetic threads were subjected to microtome cutting at a length of 200 μm according to IWTO-8-97. For the cutting step, synthetic fibers were blended with wool, as shown in **Figure 12**.

The wool is added to the synthetic yarn to fill the microtome slot completely and consequently have the correct number of synthetic filaments. The wool fibers are then removed using a hypochlorite solution. This treatment successfully eliminates the wool fiber without altering the structure of the synthetic yarns. The effect of hypochlorite on the synthetic yarn is checked using FTIR analysis.

As shown in **Figure 13** the oxidative treatment does not modify the chemical structure of the synthetic yarns since no significant changes can be seen in the absorption bands.

The presence of wool fibers can be observed using an optical microscope (OM), as shown in **Figure 14**.

The wool fibers can be easily recognized using MO analysis, as shown in **Figure 15**. Indeed, they have an irregular diameter and a surface structure consisting of overlapping scales. On the contrary, synthetic fibers typically have a wider diameter and a regular shape with a homogenous and smooth surface.

Figure 15 shows an example of wool fibers used during the cutting stage.

After the hypochlorite treatment, the synthetic fibers were placed in an Erlenmeyer flask. For each polymeric yarn, three suspensions at 300, 500 and 900 ml were prepared and then filtered using silicon filters. The microfilaments collected on the filters were counted and the average value and standard deviation of 5 replicas were calculated.

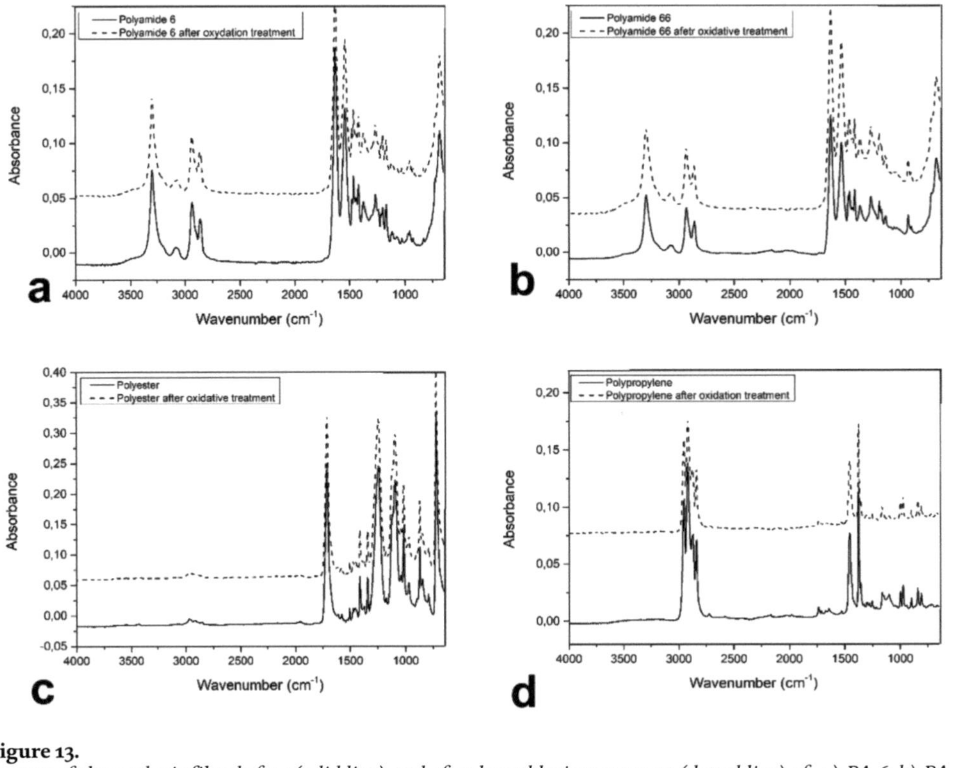

Figure 13.
Spectra of the synthetic fiber before (solid line) and after hypochlorite treatment (dotted line) of a) PA 6; b) PA 66; c) PET; d) f PP. No significant differences can be seen.

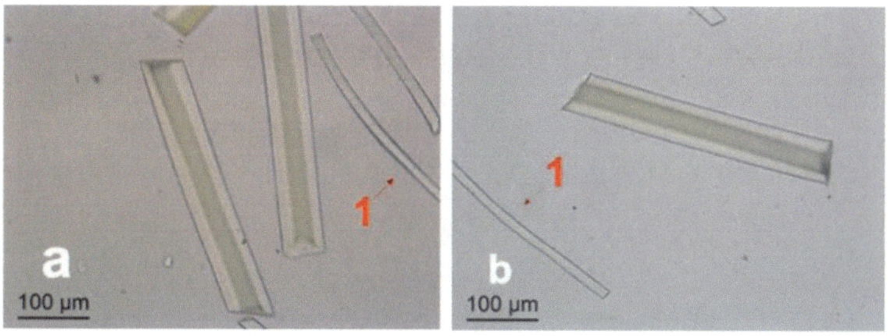

Figure 14.
Optical microscopy images (200x) of synthetic fibers (e.g., PA 6) and wool (1) cut with a microtome to 200 µm.

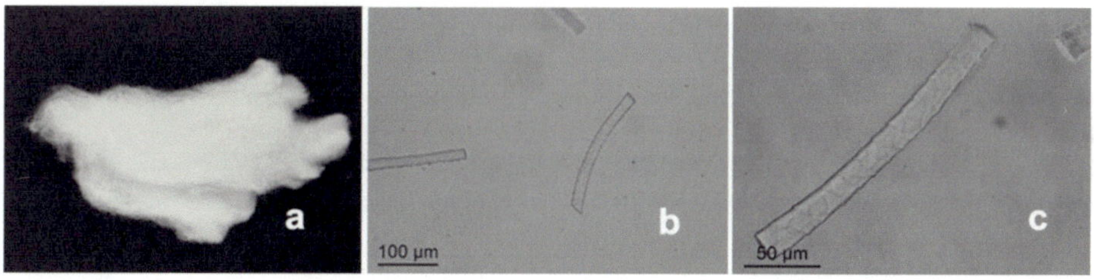

Figure 15.
a) Example of wool fine fiber used for the sample cutting stage; b) optical microscopy image of wool at 200X; c) optical microscopy image of wool at 500X. Average diameter: 16,2 µm.

An optical microscope associated with a micro-FITR was used to count the microfilaments on the filters. This technique has several advantages:

It is fast, non-destructive, reproducible, and able to collect IR signals at a high spatial resolution. Furthermore, the coupling of a MicroFTIR with an OM opens the possibility of visualization and mapping samples across the entire surface exposed.

The MicroFTIR has become an increasingly popular instrument for characterizing samples with very small dimensions which are difficult to be chemically analyzed using the conventional FTIR.

Indeed, the microscopic component provides information about morphology, size, color, and shape. On the other hand, the spectroscopic component provides information about the specific chemical bonds by capturing the absorption spectrum of the microplastic, thus performing qualitative analysis. Finally, the possibility of developing an automated spectroscopic analysis procedure is more efficient and labor-saving than other analytical methods. In MicroFTIR mapping mode, it is possible to collect spectra in different sampling points that are measured and integrated and then used to map the distribution, as shown in **Figure 16** [29].

This technology also allows the determination of the presence of contaminants inside the sample. For instance, some cellulosic fibers were found in the control water sample (hypochlorite, wool and demineralized water). Through OM analysis, the typical ribbon shape was recognized and MicroFTIR identified the characteristic absorption bands related to cellulosic fibers, as shown in **Figure 17**.

All the collected data were statistically elaborated using a logit regression analysis to study the relationship between the concentration and probability of detection of an individual microfilament, as well as the impact of the type of polymer used as shown in **Figure 18** [30]. It is as well used to investigate the relationship between a binary response variable and some other explanatory ones.

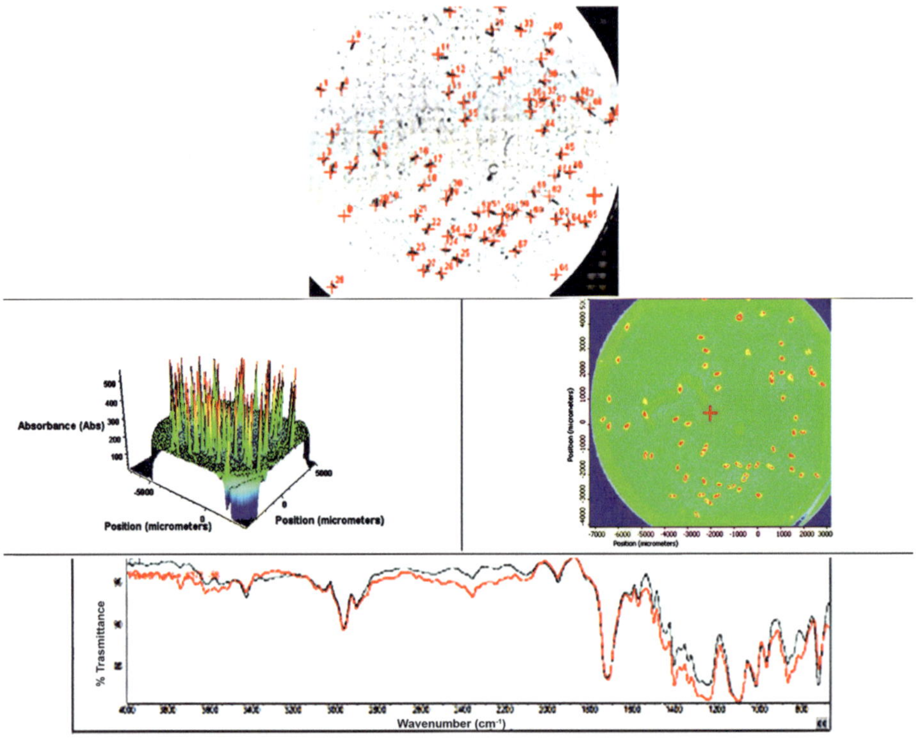

Figure 16.
Counting and chemical mapping of the microfilaments (PET) collected on a silicon filter using MicroFTIR.

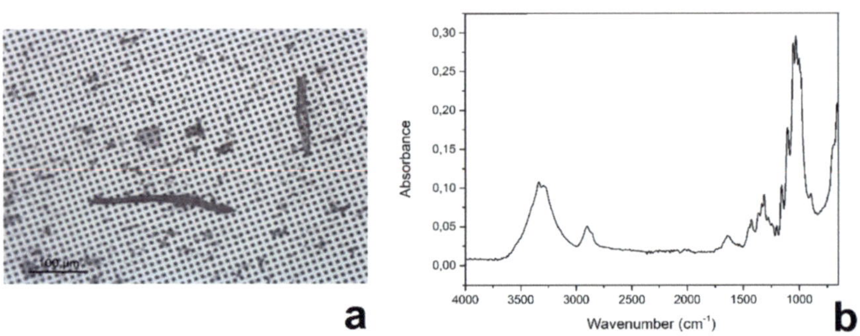

Figure 17.
a) Optical image and b) spectra of cellulosic contaminants fibers collected in a control water sample.

It was chosen because of the binary nature of the data, in which a dependent variable has two possible values expressed as identification or non-identification for each individual microfilament in the suspension. Let Y_{ij}, i = 1,..., n, j = 1,... m, denote the response, that is the number of detected microfilaments for the i-th sample and j-th replication. Let K be the theoretical number of microfilaments in the sample, that is the number of independent trials that can be performed on it. Then Y_{ij} is distributed as a binomial random variable of size K and probability of identification p_i. The logit model used explicitly the relationship between the probability of detection of the single microfilaments, p_i, and the covariates by modeling:

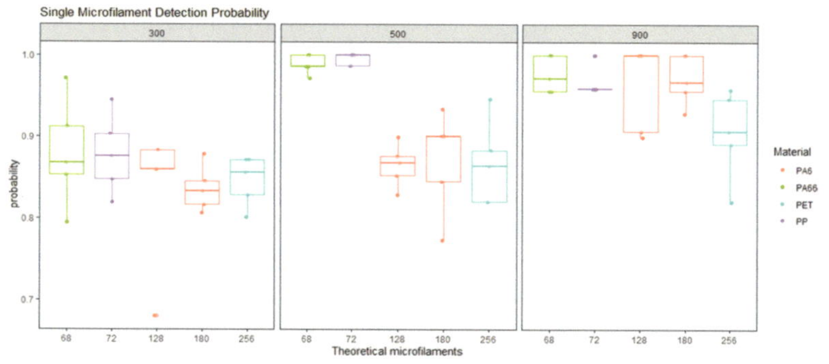

Figure 18.
Boxplot of the fraction of counted versus theoretical burrs in relation to material and solution volume.

$$\text{logit}\left(E\left(Z_{ijk}|,X_{1,ij}|,X_{2,ij}\right)\right) = \log\left(p_i/(1-p_i)\right) = \beta_0 + \beta_1 X_{1,ij} + \beta_{M(i)} \qquad (1)$$

where Z_{ijk}, k = 1,…,K is a Bernoulli random variable representing the detection of the k-th microfilament in the i-th sample and j-th replication, $X_{1,ij}$ the concentration used and $\beta_{M(i)}$ the parameter representing the material's effect used for the i-th sample.

This statistical elaboration underlines that there is a strict relationship between the concentration of the microfilaments detection probability. Indeed, increasing the number of microfilaments there is a reduction of the detection probability.

The results of statistical analysis show that:

- the fraction of microplastics detected is not the same for all materials;
- the fraction of microplastics increases with the amount of solution;
- the greater the number of theoretical microfilaments, the lower the probability of detecting all filaments;
- the type of material influences the fraction of microplastics detected;
- the probability of detecting microfilaments is greater than 95% when the microfilament concentration is less than 200 N° microfilaments/L.
- Thus, if the microfilament concentration is too high, overlapping microfilaments may occur, resulting in a loss of material identification and counting. Therefore, it is necessary to proceed with the division into several aliquots and filtration through several filters.

4. Conclusions

This chapter has tackled the problem of microplastic release from textiles by trying to identify a suitable protocol for the preparation of standard microfilaments. Indeed, there is a growing concern about the microfilament from textiles released in the environment. Since the average textile consumption is increasing, the number of synthetic microfilaments released in particular in water is rapidly enhanced. Thus, the necessity to have a reliable method for the identification and quantification of microplastic released by textiles are becoming mandatory. For this reason, in this chapter, it has been proposed not only a complete overview of the problem of the microplastics related to the textile sector but also a novel approach for the quantification and identification of them. Therefore, this chapter describes a protocol in which some standards of different synthetic fibers have been prepared in order to introduce them in a real sample. Actually, it describes the preparation of standard suspensions with a 76–853 N filaments/L concentration range using polymer threads cut at predetermined lengths of 200 mm following IWTO-8-97 and dispersed in three solutions of 300, 500, 900 ml to obtain three different concentrations. Afterward, the solutions were filtered through a silicon filter, and the collected microfilaments were counted with optical microscopy coupled with a MicroFTIR instrument. Five replicates were carried out for each sample and the data were statistically analyzed using a logit method. The probability of detecting the microfilaments is higher than 95% when the concentration of microfilaments/L is lower than 200. Thus, these microfilaments can actually work as an internal standard and the micro-FITR can be a suitable tool for the correct identification and quantification of microplastics.

Author details

Raffaella Mossotti[1*], Giulia Dalla Fontana[1], Anastasia Anceschi[1], Enrico Gasparin[2] and Tiziano Battistini[2]

1 CNR-STIIMA, Consiglio Nazionale delle Ricerche-Istituto di Sistemi e Tecnologie Industriali Intelligenti per il Manifatturiero Avanzato, Biella, Italy

2 Aquafil Group Ind. Env. Tech. Inn, Linfano, Italy

*Address all correspondence to: raffaella.mossotti@stiima.cnr.it

References

[1] Ouellette RJ, Rawn JD. Synthetic Polymers. Boston: Elsevier; 2015. pp. 397-419

[2] Sridharan S, Kumar M, Saha M, Kirkham MB, Singh L, Bolan NS. The polymers and their additives in particulate plastics: What makes them hazardous to the Fauna? Science Total Environment. 2022;**824**:153828

[3] Arpia AA, Chen W-H, Ubando AT, Naqvi SR, Culaba AB. Microplastic degradation as a sustainable concurrent approach for producing biofuel and obliterating hazardous environmental effects: A state-of-the-art review. Journal of Hazardous Materials. 2021;**418**:126381

[4] Szymańska M, Obolewski K. Microplasticsascontaminantsinfreshwater environments: A multidisciplinary review. Ecohydrology and Hydrobiology. 2020;**20**(3):333-345

[5] Reddy AS, Nair AT. The fate of microplastics in wastewater treatment plants: An overview of source and remediation technologies. Environmental Technology and Innovation. 2022;**28**:102815

[6] Periyasamy AP, Tehrani-Bagha A. A review on microplastic emission from textile materials and its reduction techniques. Polymer Degradation Stability. 2022;**199**:109901

[7] Kwon HJ, Hidayaturrahman H, Peera SG, Lee TG. Elimination of microplastics at different stages in wastewater treatment plants. Watermark. 2022;**14**(15):2404

[8] Hartline NL, Bruce NJ, Karba SN, Ruff EO, Sonar SU, Holden PA. Microfiber masses recovered from conventional machine washing of new or aged garments. Environmental Science & Technology. 2016;**50**(21):11532-11538

[9] Thompson RC, Olsen Y, Mitchell RP, Davis A, Rowland SJ, John AWG, et al. Lost at sea: Where is all the plastic? Science. 2004;**304**(5672):838

[10] Arthur C, Baker J, Bamford H. Proceedings of the International Research Workshop on the Occurrence, Effects , and Fate of Microplastic Marine Debris. Tacoma, WA, USA. 2009. p. 530

[11] Cole M, Lindeque P, Halsband C, Galloway TS. Microplastics as contaminants in the marine environment: A review. Marine Pollution Bulletin. 2011;**62**(12):2588-2597

[12] European Chemicals Agency. Committee for Risk Assessment (RAC) Committee for Socio-Economic Analysis (SEAC). Background Document. 2018;**1**:385

[13] Halbeisen M. Textiles. Oxford: Elsevier; 2005. pp. 1-8

[14] Greaves PH, Saville BP. Microscopy of Textile Fibres. Garland Science, London, UK. 1995

[15] Běhal J, Valentino M, Miccio L, Bianco V, Itri S, Mossotti R, et al. Toward an all-optical fingerprint of synthetic and natural microplastic Fibers by polarization-sensitive holographic microscopy. ACS Photonics. 2022;**9**(2):694-705. DOI: 10.1021/acsphotonics.1c01781

[16] The Fiber Year. 2021

[17] A New Textiles Economy: Redesigning Fashion's Future.

Ellen MacArthur Found; 2017. pp. 1-150. Available from: https:// ellenmacarthurfoundation.org

[18] De Falco F, Di Pace E, Cocca M, Avella M. The contribution of washing processes of synthetic clothes to microplastic pollution. Scientific Reports. 2019;**9**(1):6633

[19] Napper IE, Barrett AC, Thompson RC. The efficiency of devices intended to reduce microfibre release during clothes washing. Sciebce Total Environment. 2020;**738**:140412. DOI: 10.1016/j.scitotenv.2020.140412

[20] De Falco F, Cocca M, Avella M, Thompson RC. Microfiber release to water, via Laundering, and to air, via everyday use: A comparison between polyester clothing with differing textile parameters. Environmental Science & Technology. 2020;**54**(6):3288-3296

[21] Wright SL, Ulke J, Font A, Chan KLA, Kelly FJ. Atmospheric microplastic deposition in an urban environment and an evaluation of transport. Environment International. 2020;**136**:105411

[22] Zambrano M, Venditti R, Pawlak J, Daystar J, Ankeny M, Cheng J. The Generation And Aquatic Biodegradation Of Microfibres Produced From Laundering Fabrics. 2016. Available from: https://baumwollboerse.de/wp-content/uploads/2018/03/Venditti_R_Bremen2018-to-submit.pdf

[23] Daria M, Krzysztof L, Jakub M. Characteristics of biodegradable textiles used in environmental engineering: A comprehensive review. Journal of Cleaner Production. 2020;**268**:122129

[24] Yuan Z, Nag R, Cummins E. Human health concerns regarding microplastics in the aquatic environment - from marine to food systems. Science Total Environment. 2022;**823**:153730

[25] Carr SA, Liu J, Tesoro AG. Transport and fate of microplastic particles in wastewater treatment plants. Water Research. 2016;**91**:174-182

[26] Murray F, Cowie PR. Plastic contamination in the decapod crustacean Nephrops Norvegicus (Linnaeus, 1758). Marine Pollution Bulletin. 2011;**62**(6):1207-1217

[27] Graham ER, Thompson JT. Deposit- and suspension-Feeding Sea cucumbers (Echinodermata) ingest plastic fragments. Journal of Experimental Marine Biology and Ecology. 2009;**368**(1):22-29

[28] Mossotti R, Dalla Fontana G, Anceschi A, Gasparin E, Battistini T. Preparation and analysis of standards containing microfilaments/microplastic with fibre shape. Chemosphere. 2021;**270**:129410

[29] Hospodarova V, Singovszka E, Stevulova N. Characterization of cellulosic Fibers by FTIR spectroscopy for their further implementation to building materials. American Journal of Analytical Chemistry. 2018;**09**(06):303-310

[30] Claessens M, Van Cauwenberghe L, Vandegehuchte MB, Janssen CR. New techniques for the detection of microplastics in sediments and field collected organisms. Marine Pollution Bulletin. 2013;**70**(1-2):227-233

Chapter 4

Round Robin Test on Microplastic Counting and Identification Method

Raffaella Mossotti, Giulia Dalla Fontana, Anastasia Anceschi, Enrico Gasparin and Tiziano Battistini

Abstract

The aim of this work is to verify the quality, robustness, and accuracy of a standard analytical protocol for the determination of microplastics in aqueous textile matrices. In order to reach this objective, a round robin scale identification and quantification test program was conducted. In particular, this chapter describes the round robin test, an interlaboratory comparison test on standard microfilament suspensions initiated in November 2021 by an expression of interest open call. In total, 18 laboratories expressed their interest, and 13 participants sent their results. Each of these laboratories received a set of 10 samples, accompanied by a protocol. The 10 samples consisted of three replicates per type of three different synthetic yarns and a control sample. The data required were the number of microplastics per sample recognized as fibers or particles, microplastic fiber lengths and diameters, and identification of the polymer using vibrational spectroscopy (μ-FTIR and/or μ-Raman). The data collected were statistically elaborated. The results highlighted that the laboratories had different recovery rates directly related to their specific procedures and equipment. Although there were issues related to the correct use of the standard method and to the behavior of operators, the method proved to be valid for the determination of microplastics in aqueous matrices.

Keywords: standard microplastic suspensions, interlaboratory test, quantification of microplastics, vibrational spectroscopy, microplastic counting

1. Introduction

Microplastics are considered to be emerging pollutants in aquatic and terrestrial environments. Microplastics are generally defined as particles with dimensions in the range of 5 mm, and this term denotes microscopic plastic particles such as fragments, beads, or fibers [1]. They have been detected worldwide and are currently present even in remote areas such as Antarctica [2].

In recent years, a particular kind of microplastic has gained the attention of researchers and scientists after its problematic occurrence in the water environment. Indeed, among the different microplastic forms, studies have demonstrated that the predominant shape is the fibrous or the filament form in marine and freshwater ecosystems [3, 4]. Microplastic fibers, also called microfilaments, can be produced by the fragmentation of large plastics, in particular from textile garments [5]. Many synthetic and natural microplastics are released from textiles during domestic or industrial washing processes [6]. Out of all the released microfilaments found in the environment, the synthetic ones play a crucial role as pollutants of the environment. Polyesters (PET), acrylics (PAN), and polyamides (PA) are the major contributors [7]. Thus, all the source of microplastic in water are summarized in **Figure 1**. Synthetic microplastics are released from different sources, such as personal care products, city dust, and textiles [8]. Specifically, textiles can be a source of microfilaments during their production, use, and disposal stages. The mechanical abrasion and the physical stress applied to garments in any life stages are responsible for the shedding of microplastics with a fibrous shape [9].

Furthermore, domestic filters and wastewater treatment plants (WWTPs) are sometimes unable to trap them totally. It has been estimated that significant numbers of microfilaments escape from the traps and up to 40% can enter rivers, lakes, and oceans downstream [10]. Moreover, the sludge removed from treatment plants is usually stored or landfilled, allowing microfilaments to reach the environment again.

Despite the concerns about microplastics, the consumption of synthetic textiles is constantly growing, mainly due to fast-fashion trends. For instance, more than 45 million tons of polyester garments are produced every year [10]. Consequently, the world-released microplastic keeps rising, especially in marine and freshwater ecosystems [11]. Fortunately, several scientists worldwide have been moving in this direction in recent years to identify and limit microplastic pollution.

In particular, the occurrence of microfilaments in the aqueous environment is causing increasingly colossal concern. For this reason, in order to have a clearer view of microfilament pollution, a brief overview of the potential identification method is proposed.

1.1 Overview of qualitative and quantitative identification methods

In literature, different approaches for the determination of microplastics in aqueous matrices are reported [12]. They are chosen according to the data to be obtained and their usefulness. Mainly, several methods are used to acquire microplastic data such as color, size, shape, composition, and chemical concentration expressed in terms

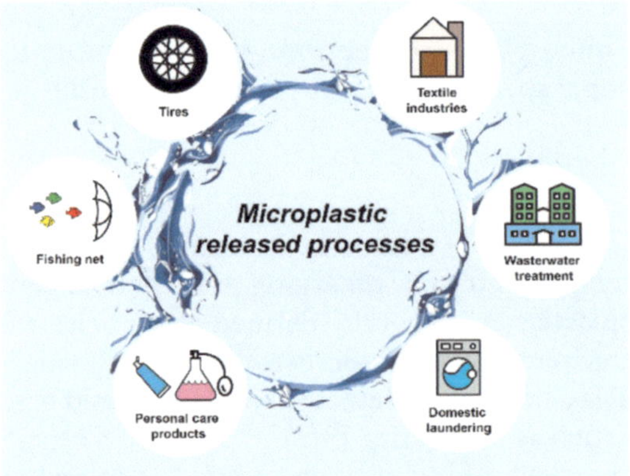

Figure 1.
Source of microplastics from textiles sources.

of number, weight, or size per unit volume or area. However, in order to obtain reliable and reproducible results, it is necessary to eliminate all possible contaminants that may interfere with the acquisition of such data.

Usually, the analysis of microplastics, as suggested by some guidelines [13], requires several steps that mainly include 3 phases: sampling, sample preparation, and determination of the type of microplastic polymer. Sample preparation is preceded by purification treatments that can be chemical or physical and are used to obtain suspensions of microplastics with reduced presence of organic and inorganic contaminants. This approach acts as a bridge between sampling and detection of microplastics as its effects influence the analytical quality of the final data in relation to specific pretreatment conditions and volumes. The identification system is applicable to any type of sample containing microplastics. An appropriate analytical approach for the identification of microplastics is shown in **Figure 2**.

2. Analytical approaches for the determination of microplastics

Microplastics are synthetic polymers of a wide variety of different shapes and colors. The choice of the analytical approach to characterize them depends on the data to be obtained in order to estimate their impact on the environment and human health. Scientific literature proposes a wide variety of techniques that provide data ranging from morphological characterization to the determination of their concentration and polymer type. The data obtained are closely related to the technique used to obtain them. Visual inspection techniques (optical and electron microscopy) are generally used for the study of morphology, color, and counting, while the study of composition is carried out by means of thermoanalytical methods, molecular spectroscopy (FT-IR and Raman), and liquid chromatography as shown in **Figure 3**.

Visual sorting is a method based on observing and counting microplastics with a stereomicroscope or optical microscope. This method allows for an error of over 70% for particles smaller than 50 μm and false positives for fibers larger than 200 μm [14, 15]. Environmental aqueous matrices are rich in cellulosic or fibrous protein material, which can be mistaken for degraded plastic material. Therefore, the lack of recognition of the chemical nature of the polymer leads to possible errors. In some cases, optical screening involves the use of dyes to increase the accuracy during visual inspection. For example, some authors identified microplastics such as PE

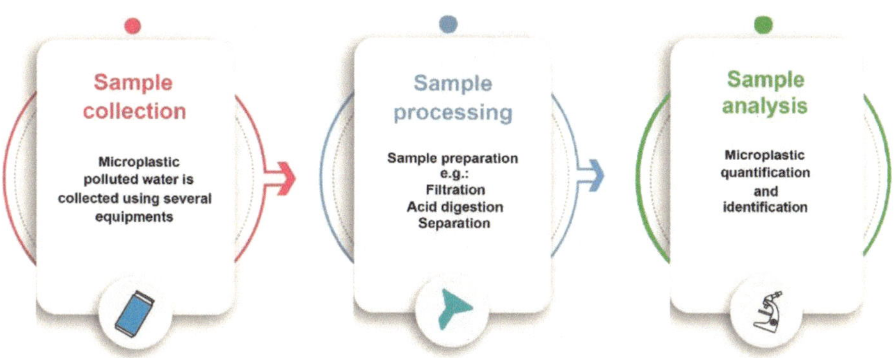

Figure 2.
Identification system for microplastics.

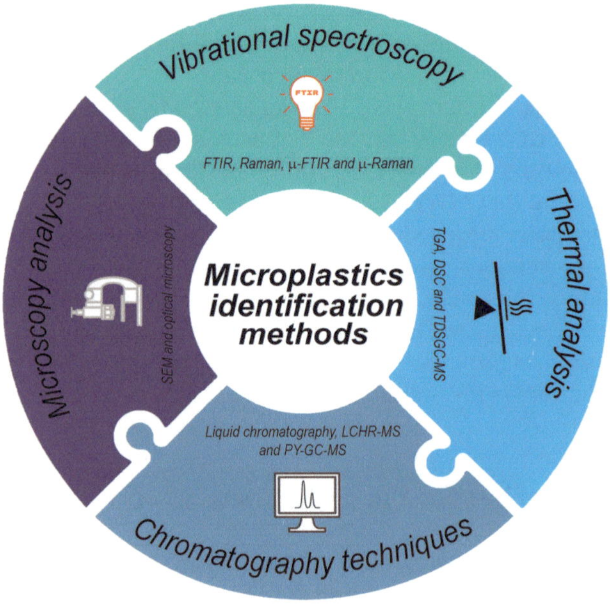

Figure 3.
Microplastic identification methods.

(polyethylene), PS (polystyrene), PP (polypropylene), and nylon 6 by using Nile Red fluorescent tagging [16, 17]. However, the coloring does not highlight polymers such as polyurethane (PU) and polycarbonate, and therefore, it limits its use.

In comparison with optical microscopy, scanning electron microscopy (SEM) can give images at high resolution of morphology, examine surface condition, and provide a qualitative determination of the chemical composition by energy dispersive x-ray spectroscopy (EDS) [18]. It is a technique coupled with SEM microscopy, generally used for the identification of organic material with high content of inorganic minerals and salts (Ca/Mg/Sr) and microplastics rich in chemical elements such as C/Cl/S/Ti.

At the same time, some authors proposed chromatography techniques coupled to high-resolution mass spectrometry (LCHRMS) for the quantification and chemical identification of microplastic (e.g., PS in natural waters). However, this technique does not provide data concerning the shape and size [9, 19].

Within visual sorting, the gravimetric method can be used to quantify microplastic particles or filaments with different sizes [20] in samples with high water volumes such as wastewater from laundry machines.

During a washing cycle of different synthetic standard fabrics or clothes at different operative conditions, the gravimetric method can be used for the determination of the mass of microfilaments released through a sieve at predeterminate porosity [6, 7, 21, 22].

At present, thermal techniques such as differential scanning calorimetry (DSC) and thermogravimetric analysis (TGA), sometimes coupled with chromatography-mass spectrometry (TDSGC-MS) or pyrolysis (Py-GC-MS), are used for the quantitative determination of microplastics [12, 23].

The samples are subjected to high-temperature treatments in order to produce thermal degradation, and the volatile compound products are analyzed for their polymer identification by means of a mass spectrometer. However, these techniques have the disadvantage of being able to analyze only samples with a size >500 μm, and this prevents the determination of microplastics. The technique requires basic sample preparation, compared to others, but it is destructive and does not allow for multiple or parallel analyses. In this field, liquid chromatography can be used, too. However, it requires high quantities of sample volumes for the preliminary extraction step of the sample. Moreover, it only provides reliable results for a limited number of polymers such as PET and PE.

The drawback of all these analytical techniques is that they work for the identification of polymer nature and other additives such as, UV or thermal stabilizers, flame retardants, dyes antioxidants, plasticizers, and so on, but they do not give any information about the physical characteristics such as shape, number of microplastics, or color [24].

Other analytical methods available for the identification of microplastic are vibrational spectroscopic techniques, such as mid-infrared (FTIR), near-infrared (NIR), and Raman spectroscopy. FTIR and Raman spectroscopy are complementary techniques that transform the interaction of light with the sample into a spectrum that contains all the information about the chemical structure [25]. Raman spectroscopy is sensitive to the variation of the polarization of the molecule during vibrational motion, while FTIR is affected by the variation of the dipole moment in reflection, transmission, and total attenuated reflection (ATR).

Raman is a fundamental technique for the recognition of aromatic compounds and double bonds, while FTIR is used for the identification of polar functional groups of molecules such as hydroxyl, carbonyls, carboxyl, amino groups, and so on. Both are nondestructive techniques; Raman can be performed on any type of matrix, even liquid; it requires minimal sample preparation and allows for the identification of contaminants of inorganic nature, too. The Raman and FTIR techniques can be coupled with an optical microscope and, at the same time, be applied for sorting and recognizing microplastics. μ-FTIR can be used for the identification of particles larger than 10 μm and μ-Raman for particles larger than 0.2 μm [26]. Moreover, SEM or AFM (Atomic Force Microscopy) in combination with infrared spectroscopy can be used to visualize and chemically identify microplastics [27]. Although many approaches can be applied for the quantification and identification of microplastic, there are no precise guidelines to follow in relation to microfilament identification. In this chapter, a standard protocol for identification of microplastic with fibrous shape is proposed by using μ-Raman and μ-FTIR, respectively, or in a complementary approach [26, 28].

2.1 Analytical approach for the determination of microplastic with fibrous shape

The analytical approach for the determination of microplastics are summarized in **Figure 4**. Typically, it involves three main stages. The first step is related to the sampling of microplastics coming from wastewater; the second step is related to the sample preparation and the third to the choice of an appropriate detection method to identify all their characteristics.

The standard protocol proposed is used to carry out qualitative and quantitative analyses of fibrous microplastics in textile aqueous matrices by means of analytical vibrational spectroscopy techniques (μ-FTIR and μ-Raman).

Textile wastewater is a complex matrix because it is rich in organic material, microplastics with fibrous shape, dyes, salt, fiber contaminants (natural fiber), and activated charcoal from the production or finishing processes of the fiber. In order to obtain information on the nature of the polymer constituting the microplastics,

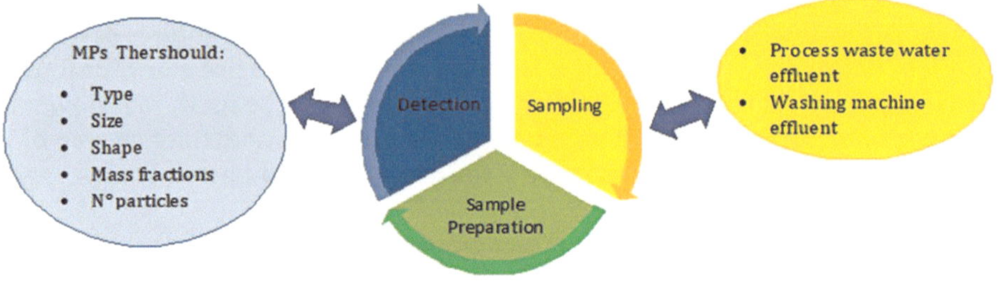

Figure 4.
Identification system for microplastics in the textile sector.

particularly when spectroscopic techniques are used, it is necessary to reduce any interference from other substances. In this regard, the analytical protocol provides suitable information on the reduction of contaminants during the sample preparation.

The following steps have been identified:

- Preliminary identification in the sample of chemical-physical parameters such as conductivity, chemical oxygen demand, and total suspended solid (TSS);
- Optical pre-screening of samples;
- Pre-treatment according to preliminary results with oxidation/digestion, acid/basic treatment, sonication, and so on;
- Filtration of microplastic samples through filters;
- Characterization of microplastics: counting and identification by optical microscopy and molecular spectroscopy.

In addition, a protocol for the preparation of standard microfilament suspensions was also prepared to facilitate the control of the whole laboratory tests (counting, monitoring, and identification of microplastics). This aspect is an added value of the method because standard fibrous microplastics with established dimensional and structural parameters are not available in the market. For this purpose, four standard suspensions of the most commonly used synthetic fibers such as PA6, PA 6.6, PP, and PET were prepared [29].

In conformity with ISO 5725 and ASTM E691 standard procedures, the protocol was validated by means of a round robin test (RRT). However, during the preparation, it was not possible to use samples from textile wastewater obtained from production processes or washing machines as they are not homogeneous and highly variable.

In order to reduce their variability, 3 replicates of 3 standard water suspensions of PA 6, PET, and PP at different concentrations were used.

3. Round Robin test

3.1 What is a RRT?

The identification and counting of microplastics face challenges due to the complexity and heterogeneity of the materials. These challenges are also paired with the lack of referenced certified methods and standards for microplastic analysis. In the absence of accredited procedures and standards, the robin round test (RRT) is one of the best approaches to identify and quantify microplastics.

The RRT is an advantageous approach since it is specific, pre-defined, and requires the involvement of several labs worldwide. It involves the presence of an organization that is able to supply samples and precise instructions and consecutively to evaluate the lab results [12]. Thus, RRT allows for the determination of the reproducibility of a process or analysis by means of multiple analyses performed independently. The statistical elaboration of the results provides a top-down evaluation of the variability by analyzing the outcomes of different labs [30]. Thus, the benefits and applicability of the RRT are diverse and can be strategically designed for various purposes.

In microplastic analysis, few studies carry out the RRT for the comparison of the results. For instance, Muller et al. presented the results of an international comparative study on the common analytical technique used in microplastic analysis [31]. Since there is a large discrepancy between the 17 labs involved in the study, the study pointed out the urgent necessity of a standardized method for microplastic analysis.

A novel approach to microplastic analysis was proposed by Mossotti et al., who found out an optimized protocol for the determination of microplastic with fibrous shape in water [29]. Three different synthetic filaments cut at predetermined lengths can be used as internal standards for microplastic identification. For the first time, an analytical method for microplastic identification was subjected to a RRT involving 18 labs around the world.

3.2 Aim of STANDARD METHOD prEN ISO 4484-2 PROTOCOL

The purpose of this RRT is to identify with adequate accuracy the number and type of standard microfilaments in suspension.

All the analyzed data of the samples were divided into polymer groups, counted, and then compared with their targets, thus obtaining the evaluation of the accuracy and reliability of the method.

3.3 Round Robin test design

3.3.1 General information

The RRT on prEN ISO 4484-2 was carried out from January to March 2022, and the trial was organized by Aquafil S.p.A (Italy) and CNR-STIIMA (Italy).

18 laboratories based in 17 European countries and a non-European one took part in the RRT. The study participants were from Italy, Germany, the U.K, Sweden, Spain, and South Korea. In **Figure 5**, the number of participants from each country is reported.

The laboratories that participated in this study represent universities, research institutions, laboratory equipment suppliers, and laboratories owned by private companies, as reported in **Figure 6**.

10 samples divided into 3 sample types called Sample 1, Sample 2, and Sample 3 (**Table 1**) were prepared for each laboratory, and in addition, 3 replicates per sample were prepared. The tenth sample corresponded to the Control Sample.

Considering the RRT membership of 18 laboratories, a total of 180 samples were prepared. The preparation of the samples, in particular the cutting of the microfilaments, is described in detail elsewhere.

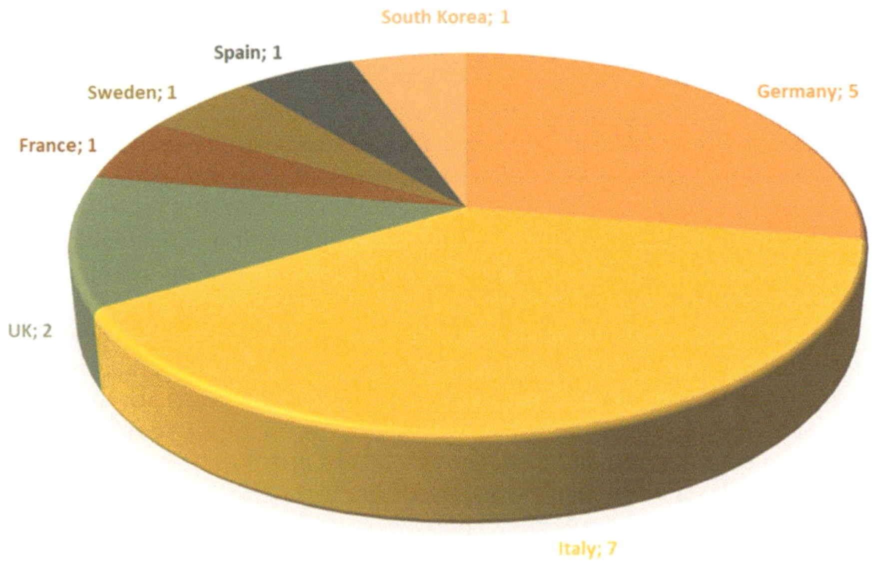

Figure 5.
Laboratories involved from different countries.

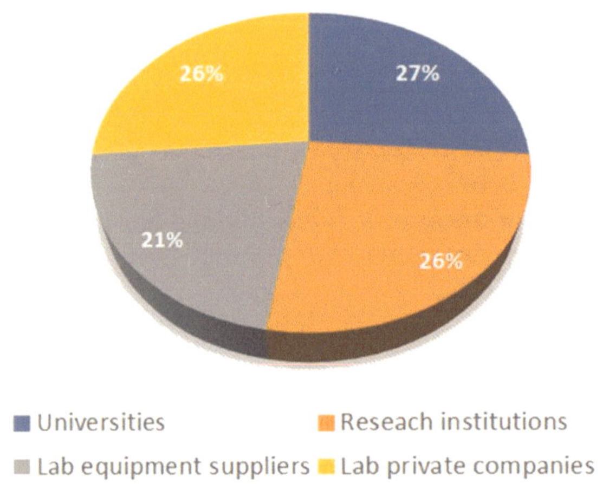

Figure 6.
The participants came from various sectors, including universities, research institutions, laboratory equipment suppliers, and laboratories owned by private companies.

After cutting, the procedure follows these steps:

a. The fibere are dispersed in 10 ml of demineralized water and 7 ml of sodium hypochlorite to remove the wool used for the microtome cutting. * ISO 1833-4:2017, Textiles-quantitative chemical analysis-Part 4: Mixture of certain protein fibers with certain other fibers.

b. The suspension was shaken in a 50 ml flask with a mechanical stirrer at 130 r.p. m. for 40 min at room temperature.

SAMPLES	STANDARD MF (microfilaments)
SAMPLE 1	MF 1 Yellow PA 6 thread (180 filaments; 3450 dtex).
SAMPLE 2	MF 2 White PP thread (80 filaments; 1300 dtex).
SAMPLE 3	MF 3 Cream PET thread (256 filaments; 2970 dtex).

Table 1.
Microfilaments used for the preparation of the standard samples.

c. The vial was washed with 50 ml aliquots of demineralized water and subsequently with a 40 ml aliquot of H_2O/ethanol (1:1) in order to recover any fibers left attached to the walls and transfer them to the bottle.

d. The sample was filled with demineralized water to a volume of 500 ml.

The described procedure is reported in **Figures 7** and **8**.

Each participating laboratory was provided with the analytical protocol (see Annex) to be followed in order to perform the determination of the microfilaments contained in the standard samples, as well as instructions for the results. For each

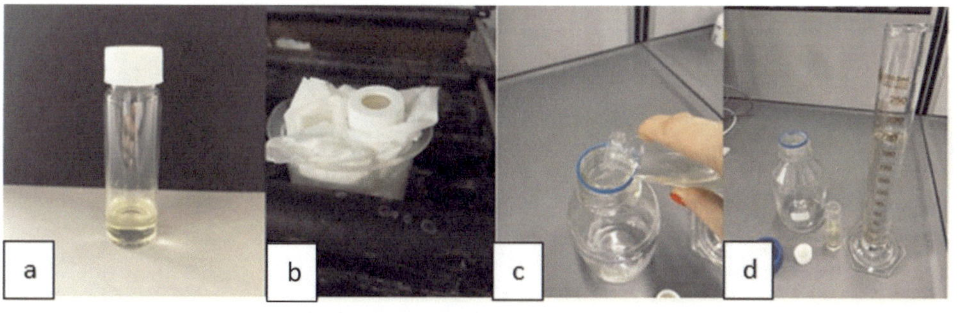

Figure 7.
Procedure for the preparation of standard sample.

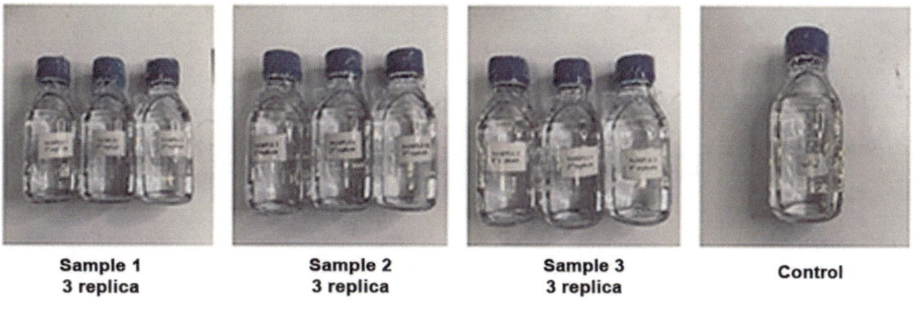

Figure 8.
A total of 10 samples were prepared for each lab.

sample, the number of total microplastics present in the sample, polymer type, and physical characteristics (microfilament lengths) had to be reported. In addition, the presence of contaminants had to be described.

4. Data collection and analysis

4.1 Working conditions

Table 2 shows the working conditions used by the laboratories when available. As shown in **Table 2**, the main drawbacks reported by laboratories are:

- Two labs used LDIR equipment (laser direct infrared imaging).
- One lab used Sterlitech silver filters.
- One lab decided to use only one filter per replicate, which made it difficult to filtrate the water sample and subsequently to count the microfilaments.

Lab.	Instrument	Filter material	Filter dimensions
Lab1	Micro-FTIR iN 10 Thermo	Silicon	Diameter 13 mm
			Pore size 1 μm
Lab2	Micro-FTIR: reflection	Cellulose acetate nitrate	Diameter 47 mm
			Pore size: unknown
Lab3	Micro-Raman (Horiba XploRA Plus)	Silicon	Diameter 13 mm; pore size 5–6 μm
Lab4	Micro-FTIR: transmission (iN 10 Thermo)	Silicon	Diameter 13; pore size 5–6 μm
Lab6	Micro-FTIR (iN 10 Thermo)	Allumina	Diameter 47 mm
			Pore size 0.2 μm
Lab7	Micro-FTIR (LUMOS II, Bruker, USA)	Silicon	(1) Diameter 10 mm; pore size 17 μm
	Micro-Raman for second filtration (XploRA PLUS, Horiba, France)		(2) Diameter 10 mm; pore size 5 μm
Lab9	Micro-FTIR (micro-ATR) (PerkinElmer Spotlight 400)	Sterlitech silver	Diameter 25 mm; pore size 3 μm
Lab10	8700 LDIR	Au-coated polycarbonate	Diameter: Unknown; pore size 0.8 μm
Lab11	8700 LDIR	Polycarbonate	Diameter: Unknown; pore size 8 μm
Lab12	8700 LDIR	Metal	Diameter 47 mm; pore size 20 μm
Lab13	Micro-Raman (XploRA Plus Microspectrometer, Horiba Scientific).	Esters of Cellulose	Diameter 47 mm: pore size 0.45 μm
Lab16	Micro-FTIR (PerkinElmer Spotlight 400)	Sterlitech silver	Diameter 11 mm; pore size 3 μm
Lab17	Micro-Raman	Silicon	Diameter 13 mm; pore size 1 μm

Table 2.
Laboratory working conditions.

- One lab decided to carry out two filtration steps: the former with a 17 μm silicon filter and the latter with a 5 μm silicon filter. The former was analyzed by μ-FTIR and the latter by μ-Raman.
- One lab had difficulty in the identification and counting of the microfilaments in some replicates due to the presence of yellowish mush (probably wool not completely degraded during treatment with hypochlorite solution).
- One lab highlighted the presence of silicates and iron probably due to contamination of demineralized water.

5. Results

All the laboratories processed the data as follows:

- Total count per target polymer type,
- Normalization of the data n° detected /n° target,
- Total count and normalization of nontarget material data.

5.1 Data analysis and visualization

After the collection of the data from each participant, the results were statistically elaborated. Actually, graphical descriptive analysis was used to compare the data obtained from the laboratories. They were not always homogenous as sometimes diverse working conditions were applied, and in few cases, some data were missing or incomplete.

5.2 Available data

Table 3 shows all the missing data in red. Eight laboratories could not perform the study due to several drawbacks, such as instrumental breakdown or unavailability of filtering systems.

5.3 Rate calculation

It was required to report all the microplastic materials found in each replicate. Since many microplastics not belonging to standard microfilaments were found in the suspensions, the recovery rate was evaluated by dividing the microplastics found in each sample by the number of standard microfilaments. This number was interpreted as the fraction of the total standard microfilaments over their theoretical quantity.

5.4 Recovery rates

Figure 9 shows the fraction of standard microfilaments found in each sample (Sample 1, Sample 2, and Sample 3) and for each replicate [1–3] obtained from all the different laboratories.

	Sample 1			Sample 2			Sample 3			Control
	R1	R2	R3	R1	R2	R3	R1	R2	R3	
Lab. 1	Green	Green	Green	Green	Green	Green	Red	Green	Green	Green
Lab. 2	Green	Green	Green	Green	Green	Green	Green	Green	Green	Green
Lab. 3	Green	Green	Green	Green	Green	Green	Green	Red	Red	Green
Lab. 4	Green	Green	Green	Green	Green	Green	Green	Green	Green	Green
Lab. 7	Green	Green	Green	Green	Green	Green	Green	Green	Green	Green
Lab. 9	Green	Green	Green	Green	Green	Green	Green	Green	Green	Green
Lab. 10	Green	Red	Red	Red	Green	Red	Red	Green	Red	Green
Lab. 11	Green	Green	Green	Red	Green	Green	Green	Green	Green	Green
Lab. 13	Green	Green	Green	Green	Green	Green	Green	Green	Green	Green
Lab. 17	Red	Red	Red	Red	Red	Green	Red	Red	Red	Red

Table 3.
Data collected from each laboratory. Red (missing data).

Only the relative fraction of the target material identified was calculated, and the absolute values were not reported.

If no evident problems of contamination occurred, the fraction of standard microfilaments was just above 0.5. However, some significant differences were found between the laboratory data. Lab 1 did not obtain accurate data for Sample 3 because during the filtration of the 1st replicate of the 3rd sample, the filter broke. The data of Lab 2 were not processable for each reference sample due to the incorrect acquisition mode of the spectra of the microplastics collected on the chosen filter. Moreover, the use of an automatic analysis system and the lack of a good reference spectral library could have amplified the mistakes. Lab 4 did not find any microplastic fibers in Sample 1 for all the replicates due to problems of working conditions during micro-FTIR automatic analysis identification, as indicated by the lab itself, for example,. selection of brightness and contrast. In fact, in this case, Lab 4 obtained data only for PP particle contaminants and not for microplastic fibers. This drawback determined a lower fraction target for all other samples Labs 6, 11 (replicate 1 of sample 2), and 16 had a recovery rate above 1, which means that they found a number of standard microfilaments higher than the present quantity. Laboratory 10 performed only one replicate for each sample and did not report anything different from the target material. Laboratory 17 did not carry out the analysis on two of the three samples and reported only the data of replicate 3 of Sample 2. The lack of results did not depend on the quality of the standard samples.

Labs 3, 7, 9, 11, and 13 reached good recovery rates for all the three samples analyzed. Labs 7 and 9 reached a recovery rate of 80–90%. The results highlighted that the labs that followed all the steps in the standard protocol reached the target fraction, confirming the quality and reproducibility of the method.

Moreover, the analysis of the results highlighted that the μ-FTIR and μ-Raman techniques also allow the identification, counting, and monitoring of the pollutants and their sources of contamination.

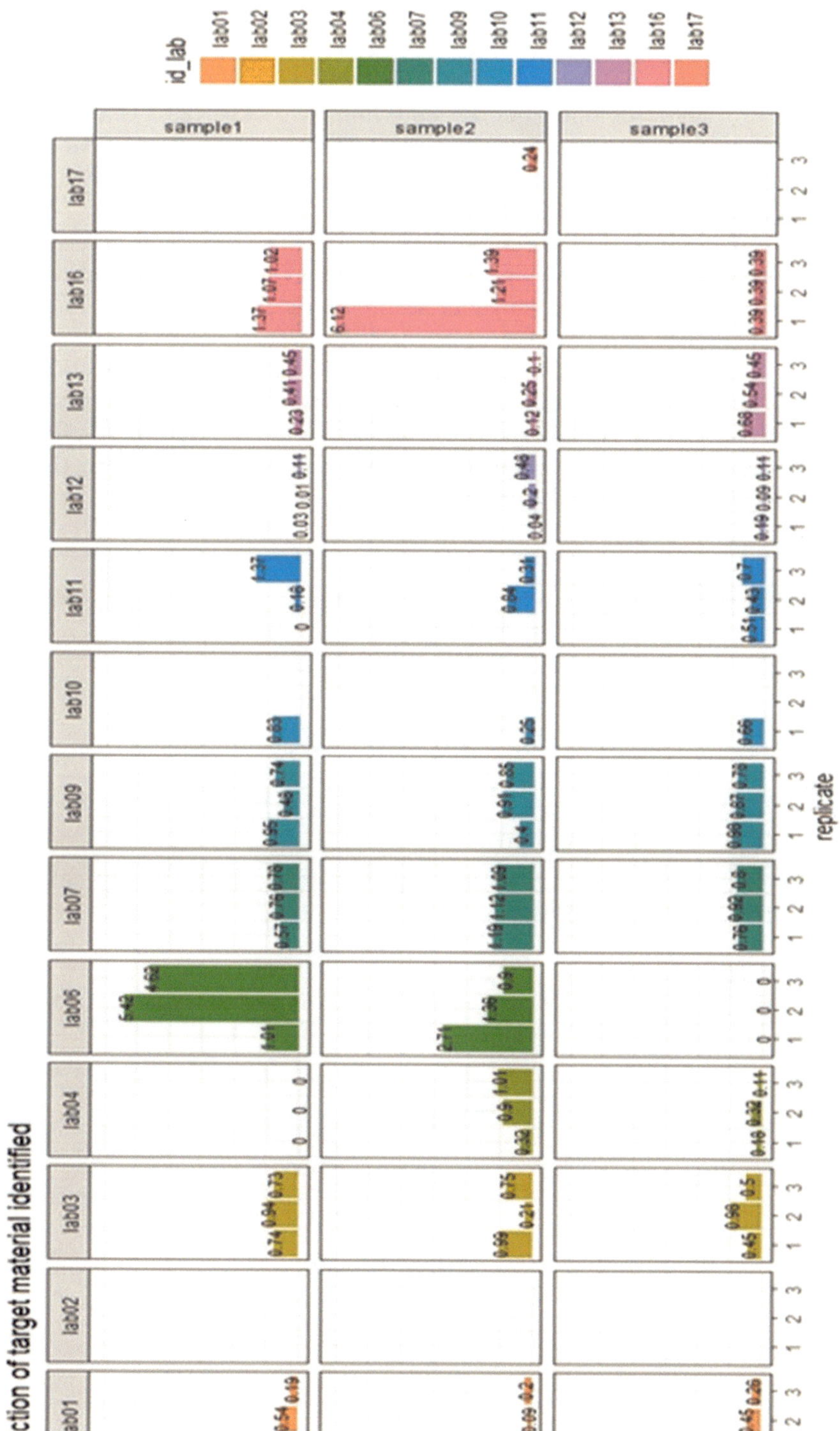

Figure 9.
Fraction of target material identified in the samples.

The main contaminants identified in analyzed samples were polypropylene (particles), natural fibers deriving from the storage system, sample preparation, and laboratory environment.

In particular, polypropylene particles were found by all the labs as a result of the containers and caps used in sample preparation. The presence of cotton fibers was identified, but it was not possible to understand whether it depended on demineralized water or dissolved wool tops.

However, when the nature of contamination was different, the source was the result of incorrect procedure. For example, one lab carried out additional sample manipulation before IR analysis, thus increasing the risk of microplastic loss and contamination (**Figure 10**).

Finally, a few participants did not perform the dimensional analysis of the microplastics identified in all the samples as suggested in the standard protocol. However, some dimensional data showed that the identified microplastic fibers were classified in the length range between 100 μm and 500 μm and diameters between 40 μm and 66 μm, confirming the control data.

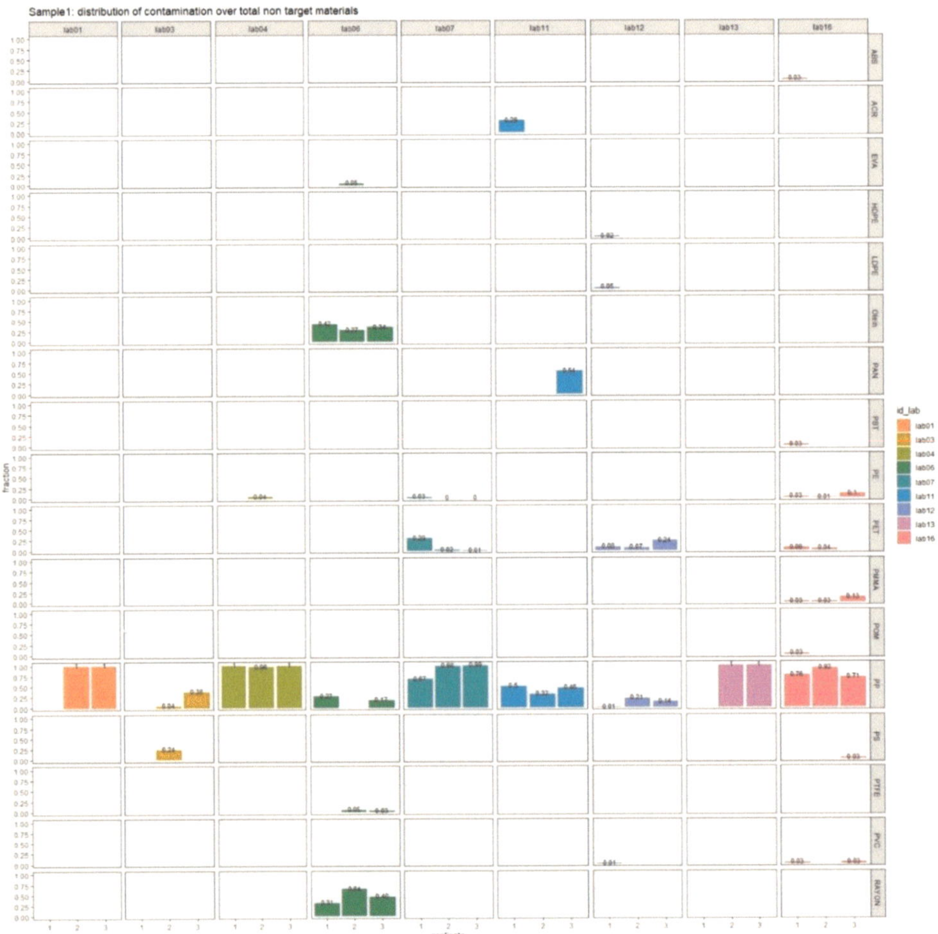

Figure 10.
Fraction of contaminant material identified in the laboratories.

6. Conclusion

This chapter has proven that the standard protocol 4484–2 allows for the determination of microplastics in aqueous textile matrices. Indeed, the round robin test was successfully carried out for an interlaboratory comparison test on the standard microfilament suspension. 18 laboratories expressed their interest, and 13 participants had sent their results. The data obtained were the number of microplastics for each sample recognized as microfilaments or particles, and identification of the polymer using vibrational spectroscopy (μ-FTIR and/or μ -Raman) had been performed. Moreover, some microfilament sizes were also indicated. The data collected were statistically elaborated by using the graphical descriptive analysis. The RRT statistical analysis has shown that the proposed protocol can be applied for the identification and counting of microfilaments in water suspension. It can be easily applied in routine analysis in order to verify the correctness of the microplastic identification procedure. Indeed, several laboratories reached a high value of recovery rate (85%), confirming the data obtained by Mossotti et al. [29]. Thus, the proposed protocol can be a suitable tool to evaluate the recovery quality of the single real sample as well as the presence of environmental contaminations.

Acknowledgements

The authors are thankful for the active support of this study by the participating laboratories that analyzed the standard samples provided, following analytical protocol, filled tables and written a test report.

Conflict of interest

The authors declare no conflict of interest.

Annex

A. STANDARD PROTOCOL (prEN ISO 4484-2)

prEN ISO 4484-2 protocol comprises:

- A.1 Filtration equipment
- A.2 Filters and their cleaning procedure
- A.3 Filtration procedure
- A.4 Results: counting, identification, and length range of microplastic with fibrous shape.
- A.4.1 Note
- A.4.2 Calculations

- A.4.3 Identification of microplastic with fiber by μ-FTIR and μ-Raman
- A.4.4 Measure of microplastic fiber lengths
- A.4.5 Test Report

A.1 Filtration equipment

The collection of microparticles with fibrous shapes is performed on one or more filters with porosity lower than the minimum diameter of the fibers used for the standard.

Equipment:

- Erlenmeyer flask containing standard solution
- Feeding funnel filter system
- Filters
- Vacuum pump
    - Filter materials: silicon, alumina, PC gold coated, cellulose acetate nitrate, or any other material with a circular or square shape.
    - The filter diameter is a choice of the lab depending on the filtration apparatus and on the spectrometer capability/availability.
    - However, in prEN ISO 4484-2 method, filter size and porosity are as follows:
    - In the case of μ-FTIR (reflection and transmission) analysis, the possible filters that can be used have a pore size not exceeding 5 μm.
    - In the case of μ Raman filters, the pore sizes that can usually be used are 0.45 μm, 0.8 μm, 1 μm, and 5 μm.
    - Vacuum pump (vacuum system filtration of 47 mm, 25 mm, 13 mm diameter, or any other diameter depending on the used filter diameter).

A.2 Cleaning procedure

- Store the filters in glass (not plastic) Petri dishes to reduce contamination from the dish itself.
- Keep filters covered whenever possible before observing the results.
- All filters must be new or clean before being used.

- Before filtration, check the filters by an optical microscope to verify that there are no interfering particles on the surface that could come from the packaging, from their handling, or from the production process itself.

- Cleaning depends on the type of filters. A physical procedure such as an ultrasonic bath or a chemical treatment by simple immersion in solvent (pure ethanol RPE) for 10 min can be used.

A.3 Filtration procedure

- Shake the bottle vigorously before filtration (possible clumps may have formed).

- Gradually filter all the stock suspension contained in the 500 ml glass bottles (50 ml at a time) through one or more filters.

- Shake the bottle each time before pouring the suspension to make sure that all the microplastic fibers on the bottle walls are removed and filtered.

- After filtering the entire stock suspension, wash the sample bottle including the cap and the funnel walls with a few ml of water/ethanol 1:1 using a glass Pasteur pipette in order to recover the possible microparticles adhering to the glass.

- Carry out the final recovery wash (of the filtering system, the gasket, and the flask containing the stock solution) with a 1:1 solution of water/ethanol and filter through Filter 2. (see below).

- The washing operations need to be repeated at least three times, each time using an aliquot of 50 ml of water/ethanol 1:1. Use an aliquot of the last wash with a glass Pasteur pipette to wash thoroughly the filter funnel, and gasket.

- One or more filters can be used to collect all the microparticles derived from all the filtering and washing procedures for each solution. Normally, we suggest to use:

- One or more filters where microparticles are collected from the filtration of the sample and from the solution of the first rinse of the flask and of the filtering system.

- One or more filter where microparticles are collected from the filtration of the subsequent rinses with the washing solution and from the solution of the subsequent thorough rinsing of the Erlenmeyer flask and all the components of the filtering system.

A.4 Results

For each replicate, the number of particles at a given volume is given by the sum of the particles collected on all the filters used for that sample (main solution and washing water).

A.4.1 Note

For more accurate results, we suggest counting the number of fibers on each section of the filter to avoid losing the count of overlapping fibers.

Zoom in and break the image into square sections to cover the entire filter, and then, manually count the microparticles in each of the sections and the number of fibers on each section of the filter.

The same procedure can be used to identify the chemical composition of microfilaments. As an alternative to manual counting, it is possible to perform automatic counting of microplastics with the software of the optical instrument.

A.4.2 Calculation

$$N°\text{filaments}\ (1)/\text{L} = \frac{N°\text{microfilaments} \times 1000}{\text{Solution volume}} \quad (1)$$

where N° filaments means number of filaments obtained by automatic or manual counting as sum of filaments counted on filters.

NOTE: samples have been prepared with an initial volume of 500 ml.

Tables (Test Report) show how the data shall be entered for each sample analyzed:

- Name of the sample
- Number of replicates
- N° of total synthetic filaments (sum of fibers counted on FILTERS), from the filtration of the whole solution, washing phase of the flask, and the filtration system with water and ethanol 1:1.
- N° of total synthetic filaments (sum of fibers counted on FILTERS), from the filtration of the whole solution, washing phase of the flask, and the filtration system with water and ethanol 1:1.

A.4.3 Identification of Fibers by μ-FTIR and μ-Raman

The operative conditions of identification and analysis of microfilaments can be optimized both with μ-FTIR and μ-Raman instrument (full mapping and/or particle by particle). FTIR can be carried out using different kinds of detectors (single detector, line array dector, focal plane array (FPA) detector).

A.4.4 Measure of microplastic fibers lengths

Moreover, the measurements of the microplastic fiber lengths and diameters can be carried out with

- Optical microscope in reflected light at 50× and 100× magnifications on a filter
- Optical microscope coupled with FTIR or Raman spectroscopy

- (*) Please define the dimensional aspect of the length in classes as in **Table 4** and also refer the diameter in microns
- 5000–1000 μm,
- <1000–500 μm,
- <500–100 μm,
- <100–50 μm,
- <50–10 μm,
- <10–5 μm,
- <5–1 μm. (only for Raman, if present),
- <1–0.1 μm (only for Raman, if present),
- (**) In this area, also report, if present, any microparticle not fibrous shaped or any presence of nonsythetic fibers

Sample 1	Replicates	N° of total synthetic filament	Identification synthetic component	Range dimensional classes (*)	Others (**)	N° filaments/L
	1°					
	2°					
	3°					
	Average					

Table 4.
Example of test report to have for each sample.

Author details

Raffaella Mossotti[1], Giulia Dalla Fontana[1], Anastasia Anceschi[1*], Enrico Gasparin[2] and Tiziano Battistini[2]

1 CNR-STIIMA, Consiglio Nazionale delle Ricerche-Istituto di Sistemi e Tecnologie Industriali Intelligenti per il Manifatturiero Avanzato, Biella, Italy

2 Aquafil Group Ind. Env. Tech. Inn, Trento, Italy

*Address all correspondence to: anastasia.anceschi@stiima.cnr.it

References

[1] Thompson RC, Olsen Y, Mitchell RP, Davis A, Rowland SJ, John AWG, et al. Lost at sea: Where is all the plastic? Science. 2004;**304**(5672):838

[2] Waller CL, Griffiths HJ, Waluda CM, Thorpe SE, Loaiza I, Moreno B, et al. Microplastics in the Antarctic marine system: An emerging area of research. Science Total Environment. 2017;**598**: 220-227. Available from: https://www.sciencedirect.com/science/article/pii/S0048969717308148

[3] Yu X, Ladewig S, Bao S, Toline CA, Whitmire S, Chow AT. Occurrence and distribution of microplastics at selected coastal sites along the southeastern United States. Science Total Environment. 2018;**613–614**: 298-305

[4] Ling SD, Sinclair M, Levi CJ, Reeves SE, Edgar GJ. Ubiquity of microplastics in coastal seafloor sediments. Marine Pollution Bulletin. 2017;**121**(1–2):104-110

[5] Gaylarde C, Baptista-Neto JA, da Fonseca EM. Plastic microfibre pollution: How important is clothes' laundering? Heliyon. 2021;7(5):e07105. Available from: https://www.sciencedirect.com/science/article/pii/S2405844021012081

[6] De Falco F, Gullo MP, Gentile G, Di Pace E, Cocca M, Gelabert L, et al. Evaluation of microplastic release caused by textile washing processes of synthetic fabrics. Environmental Pollution. 2018; **236**:916-925

[7] Dalla Fontana G, Mossotti R, Montarsolo A. Assessment of microplastics release from polyester fabrics: The impact of different washing conditions. Environmental Pollution. 2020;**264**:113960. Available from: http://www.sciencedirect.com/science/article/pii/S0269749119341065

[8] Andrady AL. Microplastics in the marine environment. Marine Pollution Bulletin. 2011;**62**(8):1596-1605. Available from: https://www.sciencedirect.com/science/article/pii/S0025326X11003055

[9] Belzagui F, Crespi M, Álvarez A, Gutiérrez-Bouzán C, Vilaseca M. Microplastics' emissions: Microfibers' detachment from textile garments. Environmental Pollution. 2019;**1**:248

[10] Acharya S, Rumi SS, Hu Y, Abidi N. Microfibers from synthetic textiles as a major source of microplastics in the environment: A review. Text Research Journal. 2021;**91**(17–18):2136-2156. Available from:. DOI: 10.1177/0040517521991244

[11] Baldwin AK, Corsi SR, Mason SA. Plastic debris in 29 Great Lakes tributaries: Relations to watershed attributes and hydrology. Environmental Science & Technology. 2016;**50**(19):10377-10385

[12] Shim WJ, Hong SH, Eo SE. Identification methods in microplastic analysis: A review. Analytical Methods. 2017;**9**(9):1384-1391

[13] GESAMP. Science for Sustainable Oceans. 2015. Available from: www.imo.org

[14] Hidalgo-Ruz V, Gutow L, Thompson RC, Thiel M. Microplastics in the marine environment: A review of the methods used for identification and quantification. Environmental Science & Technology. 2012;**46**(6):3060-3075

[15] Ivleva NP, Wiesheu AC, Niessner R. Microplastic in aquatic ecosystems.

Angewandte Chemie International Edition. 2017;**56**(7):1720-1739. DOI: 10.1002/anie.201606957

[16] Erni-Cassola G, Gibson MI, Thompson RC, Christie-Oleza JA. Lost, but found with Nile red: A novel method for detecting and quantifying small microplastics (1 mm to 20 μm) in environmental samples. Environmental Science & Technology. 2017;**51**(23): 13641-13648

[17] Maes T, Jessop R, Wellner N, Haupt K, Mayes AG. A rapid-screening approach to detect and quantify microplastics based on fluorescent tagging with Nile Red. Scientific Reports. 2017;**7**(1):44501. DOI: 10.1038/srep44501

[18] Gniadek M, Dąbrowska A. The marine nano- and microplastics characterisation by SEM-EDX: The potential of the method in comparison with various physical and chemical approaches. Marine Pollution Bulletin. 2019;**148**:210-216. Available from: https://www.sciencedirect.com/science/article/pii/S0025326X19306071

[19] Schirinzi GF, Llorca M, Seró R, Moyano E, Barceló D, Abad E, et al. Trace analysis of polystyrene microplastics in natural waters. Chemosphere. 2019;**236**:124321

[20] Lorenz C, Roscher L, Meyer MS, Hildebrandt L, Prume J, Löder MGJ, et al. Spatial distribution of microplastics in sediments and surface waters of the southern North Sea. Environmental Pollution. 2019;**252**:1719-1729. Available from: https://www.sciencedirect.com/science/article/pii/S0269749119318548

[21] Hartline NL, Bruce NJ, Karba SN, Ruff EO, Sonar SU, Holden PA. Microfiber masses recovered from conventional machine washing of new or aged garments. Environmental Science & Technology. 2016;**50**(21):11532-11538

[22] Pirc U, Vidmar M, Mozer A, Kržan A. Emissions of microplastic fibers from microfiber fleece during domestic washing. Environmental Science and Pollution Research International. 2016; **23**(21):22206

[23] Dümichen E, Barthel AK, Braun U, Bannick CG, Brand K, Jekel M, et al. Analysis of polyethylene microplastics in environmental samples, using a thermal decomposition method. Water Research. 2015;**85**:451-457

[24] Sridharan S, Kumar M, Saha M, Kirkham MB, Singh L, Bolan NS. The polymers and their additives in particulate plastics: What makes them hazardous to the fauna? Science Total Environment. 2022;**824**:153828. Available from: https://www.sciencedirect.com/science/article/pii/S0048969722009202

[25] Baruah A, Sharma A, Sharma S, Nagraik R. An insight into different microplastic detection methods. International Journal of Environmental Science Technology. 2022;**19**(6): 5721-5730. DOI: 10.1007/s13762-021-03384-1

[26] Tagg AS, Sapp M, Harrison JP, Sinclair CJ, Bradley E, Ju-Nam Y, et al. Microplastic monitoring at different stages in a wastewater treatment plant using reflectance Micro-FTIR imaging. Frontiers in Environmental Science. 2020;**8**(August):1-9

[27] Akhatova F, Ishmukhametov I, Fakhrullina G, Fakhrullin R. Nanomechanical atomic force microscopy to probe cellular microplastics uptake and distribution. International Journal of Molecular Sciences. 2022;**23**(2)

[28] Schymanski D, Goldbeck C, Humpf HU, Fürst P. Analysis of microplastics in water by micro-Raman spectroscopy: Release of plastic particles from different packaging into mineral water. Water Research. 2018;**129**: 154-162. DOI: 10.1016/j.watres.2017.11.011

[29] Mossotti R, Dalla Fontana G, Anceschi A, Gasparin E, Battistini T. Preparation and analysis of standards containing microfilaments/microplastic with fibre shape. Chemosphere. 2021 May;**270**:129410

[30] Kalaronis D, Ainali NM, Evgenidou E, Kyzas GZ, Yang X, Bikiaris DN, et al. Microscopic techniques as means for the determination of microplastics and nanoplastics in the aquatic environment: A concise review. Green Analytical Chemistry. 2022;**3**:100036. Available from: https://www.sciencedirect.com/science/article/pii/S2772577422000350

[31] Müller YK, Wernicke T, Pittroff M, Witzig CS, Storck FR, Klinger J, et al. Microplastic analysis-are we measuring the same? Results on the first global comparative study for microplastic analysis in a water sample. Analytical and Bioanalytical Chemistry. 2020; **412**(3):555-560

Chapter 5

Mechanical and Thermal Properties of HDPE/PET Microplastics, Applications, and Impact on Environment and Life

Mikail Olam

Abstract

Microplastics (MPs), which have recently threatened living organisms, are widely distributed throughout the world's fresh waters, oceans, and seas. HDPEs and PETs are produced and used in significant quantities in plastics. High-density polyethylene (HDPE) and polyethylene terephthalate (PET), which can survive in the natural environment for many years, are resistant to thermal, mechanical, and biological effects. This study examined the current developments in the sources of high-density polyethylene microplastics (mHDPE) and polyethylene terephthalate microplastics (mPET), and their disposal and properties. mHDPE and mPET microplastics consist of several sources, including their debris that breaks down their waste into smaller pieces as a result of physical and chemical processes, as well as micro-sized pieces of plastic commonly applied in personal care products or synthetic textiles. mHDPE and mPET pollution has become an important environmental problem with the potential to harm human health by entering the human and animal food chain. mHDPEs and mPETs, which enter the living organism through ingestion, inhalation, and dermal contact in general, adversely affect the cellular mechanisms in different parts of the body. In addition, they are decomposed into free radicals by the effects of external factors such as light and temperature, as well as biological agents and chemical wastes in the environment, which significantly affects the sustainability of the ecological environment.

Keywords: microplastic, waste, pollution, environment, properties, application

1. Introduction

Microplastics (MP) are defined as smaller pieces of synthetic plastic polymers, which are between 1 μm and 5 mm [1, 2]. MPs are used in many areas such as personal care products and synthetic textile products. However, MPs, also formed by the breakdown of plastic waste in the environment, are a global concern in today's world, where they are ubiquitous in all environmental environments [3]. They are commonly found in air, soil, and water [4–6]. MPs are highly discussed due to their adverse effects on the ecological environment, social economy, and human health. However, despite their negative effects on the environment, they are frequently used in fields such as medicine and industry. MPs are divided into two groups as primary and secondary [7]. The primary group is the industrial production of plastic microbeads of different sizes. These products are widely used in areas such as raw materials in the manufacturing industry, personal care products, and sandblasting media [8, 9]. The secondary MPs are formed both during the use of products and when plastic wastes are broken down into smaller sizes depending on weather conditions that are exposed to light, heat, and mechanical stress [10]. However, the properties of plastics such as thermal, mechanical, electrical, and physical may vary depending on the production processes and the raw material used [11, 12]. Therefore, their mechanical and chemical properties may vary. Zhang et al. [13] reported that the glass transition temperature (Tg), melting temperature (Tm), and tensile strength values of the polylactic acid (PLA) filament, which was purchased from the NatureWorks company, were 61°C, 153°C, and 84 MPa, respectively. Olam and Tosun [14] showed that Tg, Tm, and tensile strength values of PLA, which was purchased from the PLA Max company, were 66°C, 160°C, and 53 MPa, respectively.

Plastics are used effectively in almost all industries, including construction, packaging, textiles, transportation, education, electricity, electronics, consumer products, and industrial machinery [15, 16]. Since the use of plastics is increasing day by day, their production is also increasing [17, 18]. According to ASTM D883 80c, plastics are divided into two groups; they are thermoset plastics and thermoplastics according to their chemical and mechanical properties [19, 20]. Thermoplastics are resins that liquefy when heated and harden when cooled [21]. Thermosets, on the other hand, do not reform under heat and pressure after they are produced [22]. Thermosetting plastics are alkyds, amine, silicones, allylics, phenolics, epoxies, urethanes, and polyester [23]. Thermoplastics are polypropylene (PP), polyamide (PA), polycarbonate (PC), high-density polyethylene (HDPE), low-density polyethylene (LDPE), polybutylene terephthalate (PBT), polyethylene terephthalate (PET), polystyrene (PS), polyether ether ketone (PEEK), polyphenylene sulfide (PPS), polyetherimide (PEI), polyamide imide

Figure 1.
Distribution of the global plastics production by type in 2021 [27].

(PAI), acrylic (PAA), acrylonitrile butadiene styrene (ABS), polylactic acid (PLA), polyvinyl chloride (PVC), polyurethanes (PUR), polyvinylidene fluoride (PVDF), and polytetrafluoroethylene (Teflon) [24–26]. PP, LDPE, PVC, HDPE, and PET constituted significant proportions in the world plastic production in 2021 (**Figure 1**).

HDPEs and PETs, which have a significant usage rate today, have an important ratio among the microplastics existing in the environment. The presence of mHDPE and mPET is commonly detected by scanning electron microscopy, Fourier-transform infrared spectroscopy, X-ray photoelectron spectroscopy, and Raman spectroscopy [28, 29]. Removal of MPs is mostly done by coagulation, filtration, adsorption, oxidation, and photocatalysis methods. However, these methods are rarely used [30, 31]. There is a need to increase the prevalence of the use of the methods and to develop new methods. Due to their small size, MPs can easily enter the environment and living organisms through air and contact [32]. If these wastes are disposed of, it is obvious that it will have many negative effects on the environment and life.

2. High-density polyethylene (HDPE)

2.1 Properties of HDPE

Polyethylene is used commonly as several types that grouped by density. These are high-density polyethylene (HDPE), low-density polyethylene (LDPE), linear low-density polyethylene (LLDPE), and very low-density polyethylene (VLDPE) [33]. The mechanical properties of PE are highly dependent on variables like its crystal structure, molecular weight, and extent and type of branching. Physical and thermal properties of HDPE are given in **Table 1**; mechanical and electrical properties are given in **Table 2**. Although the O_2 permeability of HDPE is higher than most plastics such as PLA, PVC, PET, and PVDC, its water permeability is quite low [50]. The thermal conductivity of HDPE is about 0.5 W/mK at 23°C, while that of PLA, ABS, and PP is 0.2 W/mK, and PA is 0.3 W/mK [51]. However, its Tm temperature is 134°C, which is low compared to other plastics (>200°C) [52]. Although this limits HDPE's use at high temperatures, it is still a thermoplastic with adequate mechanical and thermal properties for many applications. Although its tensile strength is as good as PA, PC, and PS, it is one of the polymers with the highest impact strength value [53]. Due to HDPE's low water absorption (<0.4), high glass transition temperature (102°C) and hardness (94), satisfactory tensile strength (18 MPa), antibacterial properties, and chemical and UV resistance, it is widely used in the chemical and food packaging industry.

2.2 HDPE applications

Polymers, which are used as one of the materials in the manufacture of implants, are one of the most popular synthetic materials [54]. They are preferred because of their low price, good electrical properties, chemical inertness, and easy processing [55]. PE is also used in the medical sector, both in the production of medical equipment and in the production of implants. HDPEs are commonly used for catheters, facial restoration, acetabular endoprosthesis, ear reconstruction, nasal dorsal augmentation, mandibular contours, facial contouring, orbital floor, and socket reconstruction (**Figure 2a**) [58–63]. High-density polyethylene (HDPE) is also preferred as a low-cost alternative material to replace the lost tissue in living tissue (**Figure 2b**) [64].

Chemical formula	$(C_2H_4)_n$	
Density (g/cm^3)	0.960	[34]
Specific Heat Capacity (J/KgK)	1.33	[35]
Thermal conductivity (W/mK, 23°C)	0.45	[35]
Mw (g/mol)	150,000	[36]
Mn (g/mol)	14,600	[36]
PDI (Mw/Mn)	10.3	[36]
O_2 permeability (cm^{3}µm/m^{2}h atm)	166–3041	[37]
CO_2 permeability (cm^{3}µm/m^{2}h atm)	9979–18,215	[37]
Water absorption (%, after 24 h)	<0.4	[38, 39]
Resistance to ultraviolet	Good	[40]
Tg (°C)	102	[35]
Tc (°C)	116	[41, 42]
ΔTc (J/g)	172	[43]
ΔTm (J/g)	215	[35]
Tm (°C)	134	[41, 42]
Xc (%)	65	[44]

Table 1.
Physical and thermal properties of HDPE.

Storage modulus at 25°C (MPa)	1200	[41]
Storage modulus at 80°C (MPa)	500	[41]
Loss modulus at peak (°C)	80	[45]
Tensile strength (MPa)	15–20	[46, 47]
Young's modulus (MPa)	1095	[48]
Flexural strength (MPa)	682	[45]
Elongation at break (%)	53	[46, 47]
Flexural modulus (MPa)	904	[49]
Impact strength (kJ/m^2)	31	[49]
Hardness (Shore-A)	94	[34]
Dielectric constant (1 MHz)	2.3	[42]
Dielectric strength (kV/cm)	2.2	[42]

Table 2.
Mechanical and electrical properties of HDPE.

Thermoplastic matrix composites are used in semi-structural and engineering applications. HDPE-based hollow particle-filled composite materials can be produced using the polymer injection molding process (**Figure 3a**) [70, 71]. In addition, HDPE is used in products such as storage boxes, beverage bottles, consumer electronics

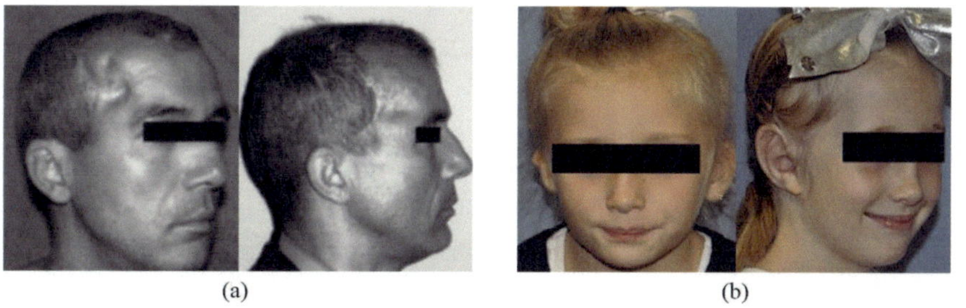

Figure 2.
Implant application of HDPE (a) MedPor implant (left: before, right: after) [56] (b) Microtia repair (left: before, right: after) [57].

Figure 3.
HDPE applications (a) Foam [65, 66] (b) Microducts [67] (c) Pipes [68] (d) Package [69].

products, and cases for automotive molded parts. HDPE with good dielectric property is used as electronic material in the form of microduct (**Table 2**) (**Figure 3b**). HDPE is used as a pipe due to its good mechanical properties such as impact strength, hardness, and tightness (**Tables 1** and **2**). HDPE pipes, unlike ferrous pipe systems, does not corrode, so there is no structure that will create internal resistance against the liquid flowing through it (**Figure 3c**). Therefore, it is preferred due to features such as long service life, high flow capacity, and ease of assembly. In addition to good barrier properties against oxygen and water vapor, packaging materials must also have antibacterial properties to preserve the physicochemical and organoleptic properties of food and beverages [72]. Since HDPE meets these features, it has an important use in the packaging industry (**Figure 3d**).

3. Polyethylene terephthalate (PET)

3.1 Properties of PET

PET is a thermoplastic with high resistance to most solvents, weak acids, and bases; strength; gloss; high impact resistance; and hardness [73]. It is also resistant to many other chemicals such as hexane, methanol, sulfur dioxide, and solutions with low acid concentrations [74–76]. The physical and chemical properties of PET are examined in **Table 3**; it is a valuable hydrocarbon containing 62% C, 4% H, and 34% O and has a high calorific value. PETs are self-extinguishing and have very low gas permeability, good adhesion and weldability, high hardness, good refractive index, and good resistance to ultraviolet [89]. They have very low O_2 and water permeability

Chemical formula	$C_{10}H_8O_4$	
Density (g/cm^3)	1.38–1.56	[77–79]
Specific Heat Capacity (J/kg K)	1000–1350	[77, 79]
Thermal conductivity (W/mK, 23°C)	0.15–0.4	[77–79]
Mw (g/mol)	30,000–80,000	[80, 81]
Mn (g/mol)	8775	[80]
PDI (Mw/Mn)	3.5–3.7	[80, 82]
Elements content (wt%)	62%C, 4%H, 34%O	[83, 84]
Lower heating values	22 MJ/kg	[85, 86]
Higher heating values	36 MJ/kg	[87]
Opacity	0.71	[88]
Flammability	Self-extinguishing	[89]
Refractive index	1.58–1. 64	[89]
Resistance to ultraviolet	Good	[89]
Freezing resistance (°C)	−50	[78]
Usable max. Temperature (°C)	70	[90]
O_2 permeability (%)	0.1–0.4	[50, 89]
CO_2 permeability (%)	0.46	[89]
Water absorption (%, after 24 h)	0.3–0.5	[79, 81]
Tg (°C)	67–80	[80, 91–93]
Tcc (°C)	115–140	[91–94]
ΔTcc (J/g)	12–34	[88, 91, 94]
Tc (°C)	194–205	[88, 91, 95]
ΔTc (J/g)	29–55	[88, 92, 95, 96]
ΔTm (J/g)	35–50	[88, 94, 96]
Tm (°C)	248–250	[91–95]
Xc (%)	13–36	[88, 91, 92]

Table 3.
Physical and thermal properties of PET.

compared to most plastics (PS, PVC, etc.) [97]. The thermal conductivity of PET is also very low compared to PLA, ABS, HDPE, PP, and PA. Although it is a fairly transparent and colorless material, it usually appears opaque and off-white as the thickness increases. Depending on the thermal and process conditions, it can be found in semi-crystalline or amorphous form. Because of this, it may appear dull, white, or glassy. It depends on process parameters such as crystal structure, processing temperature, cooling rate, and stretching. Crystallization is an important parameter that affects the properties of polymers. The benzene ring in the main chain of PET provides both slow crystallization during cooling and hardness. This adversely affects the spinning process of high-speed fibers [98]. Glass transition temperature (Tg) is an important value that affects the material properties and potential applications of a polymer [99]. Mechanical properties (deformation modulus, etc.) and physical properties (density, volume, specific heat, etc.) of polymers in glass transition state are not. Due to the Tg temperature, PET loses its glassy property and becomes viscous at 67–80°C; its cold crystallization (Tcc) is in the range of 115–140°C, melting temperature is 248–250°C, and enthalpy of melting (ΔTm) is 35–50 J/g; its hot crystallization temperature (Tc) is 194–205°C, and its enthalpy (ΔTc) is 29–55 J/g (**Table 3**) [100]. PETs begin to thermally degrade at temperatures above 340°C [101]. Although the melting temperature (Tm) of PET is not as high as PEEK (334°C), it is higher than that of PP (170°C), LDPE (134°C), PS (106°C), and PVC (199°C) [102].

PETs have excellent mechanical properties, creep resistance, fatigue resistance, friction resistance, and dimensional stability over wide temperature ranges (**Table 4**) [99, 110]. However, the disadvantages of slow crystallization rate, machining difficulties, high molding temperature, and poor impact

Storage modulus at 25°C (MPa)	2000–4200	[91, 103]
Storage modulus at 80°C (MPa)	242	[91, 103]
Loss modulus at peak (°C)	65–80	[91, 103]
Tensile strength (MPa)	40–60	[91, 104]
Young's modulus (MPa)	1000–3500	[91, 104, 105]
Flexural strength (MPa)	55–100	[91, 104, 105]
Elongation at break (%)	19–46	[104, 106, 107]
Flexural modulus (MPa)	2000–3500	[91, 103, 105, 107]
Impact strength (kJ/m^2)	4.6	[105]
Hardness (Shore-A)	96	[105]
Tan δ at peak (°C)	75–100	[91, 108]
Tan δ values at peak	0.42	[91]
Dielectric constant (1 MHz)	3.0–3.5	[78, 93, 109]
Dielectric strength (kV/cm)	150–200	[78, 93, 109]
Dissipation factor (1 kHz)	0.002	[78, 93, 109]
Surface resistivity (Ohm/sq)	1013	[78, 93, 109]
Volume resistivity (Ohm/cm)	>1014	[78, 93, 109]

Table 4.
Mechanical and electrical properties of PET.

performance limit their use [111]. The storage modulus of PET, which is the plastic deformation energy of a polymeric material, is 2000–4200 MPa at 25°C and 242 MPa at 80°C [112]. Tensile strength, flexural strength, Young's modulus, elongation at break, impact strength, flexural modulus, and hardness of PET are approximately 40–60 MPa, 55–100 MPa, 1000–3500 MPa, 19–46 MPa, 4.6 kJ/m^2, 2000–3500 MPa, and 96, respectively (**Table 4**). PET's tensile strength (26 MPa) at 77 K is higher than PA (14.5 MPa), PC (13.5 MPa), Teflon (4.3 MPa), and PVC (9.5 MPa), and its elongation at break value is higher than theirs [113].

Dielectric is an important property of insulating materials. When an ever-increasing voltage is applied to an insulating material, it eventually reaches a point where its electrical properties deteriorate, causing a drop in resistance, and it begins to lose dielectric strength. PET shows a dielectric property of 150–200 kV/cm (**Table 4**).

3.2 PET applications

PET has wide use in many fields (packaging, textile, medical, etc.) due to its satisfactory properties [114–116]. In fact, PETs that have completed their useful life (waste) are ground to micron sizes and used as concrete additives [117]. Micro PETs are generally produced and used in spherical and fibrous structures [118]. They are used in protective fabrics, filters, wound dressings, drug delivery, and scaffolds [119–121]. Because of PET's properties such as biocompatibility, high uniformity, mechanical strength, and chemical resistance, it has been successfully used in vascular prostheses for large vascular grafts and in various surface modification methods to improve the cell adhesion properties (**Figure 4a**) [127]. PETs are also used in a variety of biomedical applications such as implants, heart valves, sutures, scaffolds, surgical nets, and urinary and blood circulation catheters [122, 128]. Micro PET fibers are the most widely used synthetic fibers in textile yarn production, and their consumption is expected to be 50 Mton/year by 2050 (**Figure 4b**) [129, 130]. However, although PET's usage has become widespread in many special applications (**Figure 4a** and **b**), it is mostly used in food and beverage packaging (**Figure 4c** and **d**) [131].

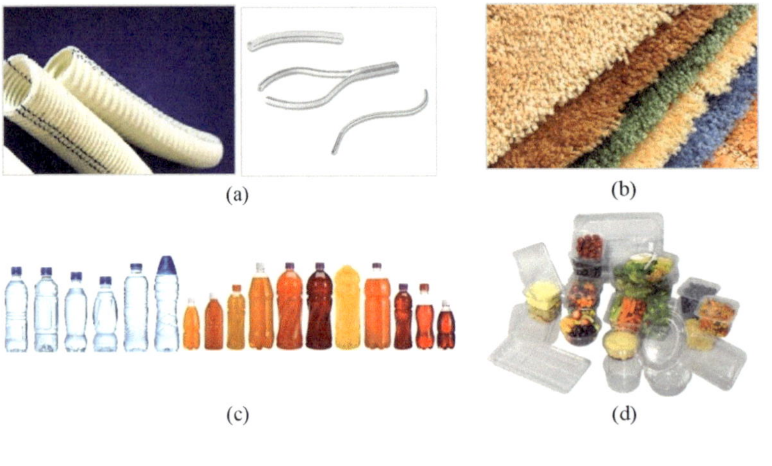

Figure 4.
PET applications (a) vascular prostheses [122, 123] (b) carpet [124] (c) beverage bottles [125] (d) food packaging [126].

4. The formation of mHDPE and mPET in nature

HDPE microplastics (mHDPE) and PET microplastics (mPET) are discharged into the environment in two ways [132]. The first of these is that they are fabricated by producers in <5 mm dimensions in the form of granules, powders, or pellets for later use in applications [133]. During the production process and during their use in the application areas, they are discharged into the environment through the disposal in air, wastewater, and other waste. The second is formed by the degradation of HDPEs and PETs under environmental conditions, which have reached the end of their useful life and are collected directly into the environment as waste or in municipal waste collection areas [134]. In nature, HDPE and PETs are resistant to chemical degradation and take decades for theenvironmental residues to decompose completely [135]. This degradation is the reduction of molecular weight as a result of chemical changes in the structure of the polymer [136]. Waste HDPE and PET exposed to sunlight undergo photooxidation as a result of the absorption of high-energy wavelengths of the ultraviolet (UV) spectrum [137]. As long as decomposition continues in the presence of oxygen, temperature-dependent thermo-oxidative reactions may occur. In addition, degradation may occur due to both biological and mechanical stresses. As a result, these decomposed HDPE and PET wastes become brittle and gradually break up into smaller pieces of micron size [138].

Photodegradation is one of the common degradation processes of polymers, which provokes cross-linking and chain scission reactions [139]. In the photooxidation mechanism that occurs during the UV-irradiation period, first carbonyl groups are formed; then, vinyl and hydroxyl/hydroxyperoxide groups are formed, and these chemical changes can be observed by FTIR spectroscopy [140, 141]. In the FTIR spectra of HDPE, its peaks are occurred at 3300–3600 cm^{-1} of hydroxyl groups (−OH), 1700–1800 cm^{-1} of carbonyl groups (>C=O), and 1600–1650, 989 and 908 cm^{-1} of vinyl groups (−C=C−) [142, 143]. The effect of UV irradiation on polymers, carbonyl, vinyl, and hydroxyl groups is studied as an indicator of polymer basic scission [144]. Since the main photooxidation product groups of HDPE are carbonyl and vinyl, the effect of UV radiation can be examined by looking at the carbonyl index and vinyl index [145]. When HDPEs in thin film form are exposed to UV irradiation at 280 nm at a light intensity of 500 W/m^2 at 25°C and 50% constant relative humidity, the carbonyl and vinyl indexes increase over time [139, 142]. After 30 days, the carbonyl and vinyl indexes of HDPE increase significantly [61, 140, 144]. As a result, it can be said that waste HDPEs, which are exposed to UV irradiation in nature, start to decompose after 30 days and turn into smaller particles. However, the carbonyl index of PET does not change significantly over time [146]. It can be said that PET is more resistant to UV irradiation than HDPE. In short, waste PETs take more time to decompose in nature by UV irradiation compared to HDPE.

The degradation of plastics, which exist as waste in nature, by using microorganisms is of great interest. Biological agents (bacterial and fungal species) and their metabolic enzymes, which are abundant in nature, have different degradation abilities for natural and synthetic polymers [147]. The biodegradation process of HDPE in nature is quite slow [148]. Therefore, it is necessary to recover HDPE wastes by appropriate methods such as physical, chemical, and biological processes [149]. Moog et al. [150] highlighted the potential for biodegradation of waste PETs and stated that PET substrates under varying conditions have enzyme functionality, secretion, and production of recombinant proteins. Shabbir et al. [151], in their experimental study with PE, PET, and PP microplastics, stated that there was weight loss in all MPs, indicating structural, morphological, and chemical changes, and confirmed this situation with SEM and FTIR analyses. Farzi et al. [152] investigated the biodegradation of mPETs by streptomyces bacterial species at laboratory scale and stated that the degradation is slow compared to physicochemical methods, and additional physical/chemical methods should be applied to increase the degradation rate. The biodegradation of 500 m particle size HDPE under laboratory conditions is approximately 10% in 24 days [153].

Mechanical degradation of polymers is typically limited to chain scission [154]. As given in **Tables 2** and **4**, due to the good mechanical properties of PET and HDPE such as tensile strength and elongation at break, they are slow to decompose as waste by mechanical forces in nature.

However, although the mechanical, biological, and chemical degradation times of mPET and mHDPE, which exist as waste in nature, are long separately, this time becomes shorter when it is considered that they occur together. As the use of PET and HDPE increases day by day, it creates more waste in nature, and accordingly, more microplastics are formed.

4.1 The effect of mHDPE and mPET on nature and life

Plastics are the most useful synthetic polymers used in packaging industries, agriculture, household applications, and many similar applications [155]. The unpredictable use of these synthetic polymers leads to an ever-increasing accumulation of solid waste in the natural environment [156]. This causes soil and water pollution at alarming rates, affecting the natural system and creating various environmental hazards [157]. Plastics, which are resistant to environmental influences, are seen as an environmental threat. MPs, both leaching into the environment in micron sizes during the production and usage processes by the manufacturers and occurring as small-sized plastic particles by the degradation of plastics found as waste in the environment, can lead to potential ecotoxicological effects [158, 159]. In general, the densities of microplastics found in nature vary in the range of 0.8–2 gcm^{-3}. The microplastic particle has the average weight of 12.5 μg, volume of 0.011 mm^3, and density of 1.14 gcm^{-3} [160]. Therefore, MPs can enter the body through airborne inhalation and food intake, which exist in many places in atmospheric environments today. When inhaled, they may cause inflammation or other biological responses in the lung [161]. In addition, they can cause health effects such as genotoxicity, which is due to the desorption of pollutants associated with polycyclic aromatic hydrocarbons (PAH); reproductive toxicity, which is due to the plastic itself and additives (plasticizers, dyes, etc.); mutagenicity; and carcinogenicity [161, 162]. Khalid et al. [163] stated that there are various inorganic and organic chemicals absorbed on MPs and that this poses a greater threat to living things than to MPs. In addition, MPs affect some plant community structure [164]. Issac et al. [165] stated that PE (about 54%) is the most abundant microplastic floating in the ocean.

Cheng et al. [166] investigated the effect of HDPE (25 μm) and PP (13 μm) microplastics on earthworms (*Metaphire guillelmi*) using Nile red fluorescent staining and observed ingestion by earthworms exposed to HDPE and PP microplastics. PP microplastics significantly reduced bacterial diversity and changed the bacterial community structure in the soil. Bringer et al. [167] exposed oyster embryos to different sizes of mHDPE for 24 hours and observed that the mHDPEs bind to the locomotor eyelashes of oyster D-larvae, influencing the swimming activity and development of oyster D-larvae. Kim et al. [168] conducted an experimental study using the *nematode caenorhabditis elegans* and zebra fish to determine the effects of mHDPEs on human health. They stated that it affected *caenorhabditis elegans* reproduction and zebra fish larval death at a concentration of more than 200,000 particles/mL. Jemec et al. [169] observed that mpETs, formed by abrasion and washing of textiles, were ingested by daphnia magna and increased the death rate of daphnia magna in their guts. Shen et al. [170] observed that the higher the concentration of mPETs, the more pronounced was the negative effect on *Drosophila*, with reduced egg production of the female flies and lower lipid, glucose content, and starvation resistance of the male flies. Najahi et al. [170] studied the effects of mPETs on human bone marrow mesenchymal stromal cells and adipose mesenchymal stromal cells. They found that it caused an approximate 30% reduction in proliferating cells associated with the onset of senescence or an increase in apoptosis.

Existing studies on mPETs and mHDPEs show that there is a lot of evidence that these wastes have a negative impact on living life. However, this does not cover all living things. Weber et al. [171] investigated the effect of the freshwater invertebrate amphipod *Gammarus pulex* exposed to mPETs for 48 hours and observed that the survival, development, metabolism, and nutritional activity of *Gammarus pulex* did not change significantly depending on the amount of mPET and the age of *Gammarus pulex*.

5. Conclusions

Microplastics are pollutants that accumulate in large quantities in the environment with each passing day and cause significant pollution. In addition to leaking into the environment during the production and usage processes of MPs, they can be formed by the mechanical, thermal, and biological decomposition of plastics, which are discharged into the environment after the completion of their useful life. HDPE and PETs, which have significant usage and application areas among plastics, have a very long life in the natural environment due to their chemical and mechanical resistance. In the current studies, it was seen that mHDPEs and mPETs indirectly or directly affect the habitat characteristics of living things and basic ecosystem functions. mHDPEs and mPETs can affect the living organism in which they are infested directly with their function and properties, as well as with the impurities they absorb. Although mPETs and mHDPEs do not cover all living organisms in the world, they adversely affect life and the environment. However, more scientific studies are needed to predict this situation.

Conflict of interest

The authors declare no conflict of interest.

Author details

Mikail Olam
Engineering Faculty, Mechanical Engineering Department, Firat University, Elazig, Turkey

*Address all correspondence to: mikailolam@gmail.com; 152120203@firat.edu.tr

References

[1] Lebreton L, Slat B, Ferrari F, et al. Evidence that the Great Pacific Garbage Patch is rapidly accumulating plastic. Scientific Reports. 2018;**8**:4666. DOI:10.1038/s41598-018-22939-w

[2] Karim ME, Al SS, Mahmud S, et al. Microplastics pollution in Bangladesh: Current scenario and future research perspective. Chemistry and Ecology. 2019;**36**:83-99. DOI:10.1080/02757540.2019.1688309

[3] Bashir SM, Kimiko S, Mak CW, et al. Personal care and cosmetic products as a potential source of environmental contamination by microplastics in a densely populated Asian City. Frontiers in Marine Science. 2021;**8**:604. DOI:10.3389/FMARS.2021.683482/BIBTEX

[4] Yuan W, Liu X, Wang W, et al. Microplastic abundance, distribution and composition in water, sediments, and wild fish from Poyang Lake, China. Ecotoxicology and Environmental Safety. 2019;**170**:180-187. DOI:10.1016/J.ECOENV.2018.11.126

[5] Scheurer M, Bigalke M. Microplastics in Swiss floodplain soils. Environmental Science & Technology. 2018;**52**:3591-3598. DOI:10.1021/ACS.EST.7B06003/SUPPL_FILE/ES7B06003_SI_001.PDF

[6] Allen S, Allen D, Phoenix VR, et al. Atmospheric transport and deposition of microplastics in a remote mountain catchment. Nature Geoscience. 2019;**125**(12):339-344. DOI:10.1038/s41561-019-0335-5

[7] Andrady AL. The plastic in microplastics: A review. Marine Pollution Bulletin. 2017;**119**:12-22. DOI:10.1016/J.MARPOLBUL.2017.01.082

[8] Fendall LS, Sewell MA. Contributing to marine pollution by washing your face: Microplastics in facial cleansers. Marine Pollution Bulletin. 2009;**58**:1225-1228. DOI:10.1016/J.MARPOLBUL.2009.04.025

[9] Kazour M, Terki S, Rabhi K, et al. Sources of microplastics pollution in the marine environment: Importance of wastewater treatment plant and coastal landfill. Marine Pollution Bulletin. 2019;**146**:608-618. DOI:10.1016/J.MARPOLBUL.2019.06.066

[10] Browne MA, Crump P, Niven SJ, et al. Accumulation of microplastic on shorelines woldwide: Sources and sinks. Environmental Science & Technology. 2011;**45**:9175-9179. DOI:10.1021/ES201811S/ASSET/IMAGES/MEDIUM/ES-2011-01811S_0003.GIF

[11] Xu Z, Bai X, Ye Z. Removal and generation of microplastics in wastewater treatment plants: A review. Journal of Cleaner Production. 2021;**291**:125982. DOI:10.1016/J.JCLEPRO.2021.125982

[12] Hossain N, Bhuiyan MA, Pramanik BK, et al. Waste materials for wastewater treatment and waste adsorbents for biofuel and cement supplement applications: A critical review. Journal of Cleaner Production. 2020;**255**:1-13. DOI:10.1016/J.JCLEPRO.2020.120261

[13] Zhang C, Huang Y, Luo C, et al. Enhanced ductility of polylactide materials: Reactive blending with pre-hot sheared natural rubber. Journal of Polymer Research. 2013;**20**:1-9. DOI:10.1007/S10965-013-0121-9

[14] Olam M, Tosun N. Assessment of 3D printings produced in fused

deposition Modeling printer using Polylactic acid/TiO2/hydroxyapatite composite filaments. Journal of Materials Engineering and Performance. 2022;**2022**:1-12. DOI:10.1007/S11665-021-06539-W

[15] Okunola AA, Kehinde IO, Oluwaseun A, Olufiropo EA. Public and environmental health effects of plastic wastes disposal: A Review. Journal of Toxicological Risk Assessment. 2019;**5**:1-13. DOI: 10.23937/2572-4061.1510021

[16] Chen Y, Awasthi AK, Wei F, et al. Single-use plastics: Production, usage, disposal, and adverse impacts. Science Total Environment. 2021;**752**:141772. DOI:10.1016/J.SCITOTENV.2020.141772

[17] Plastics Europe, Organisations EAOPR& R 2021. Plastics - the Facts 2021. 2021

[18] Geyer R, Jambeck JR, Law KL. Production, use, and fate of all plastics ever made. Science Advances. 2017;**3**:1-5. DOI: 10.1126/SCIADV.1700782/SUPPL_FILE/1700782_SM.PDF

[19] Hernandez EDD, Reyes-Romero JR. Characteristics of polymeric materials used in medicine. In: Materials for Biomedical Engineering: Thermoset and Thermoplastic Polymers. Elsevier; 2019. pp. 479-506. DOI: 10.1016/B978-0-12-816874-5.00014-1

[20] Dhinakaran V, Surendar KV, Riyaz MSH, Ravichandran M. Review on study of thermosetting and thermoplastic materials in the automated fiber placement process. Materials Today Proceedings. 2020;**27**:812-815. DOI:10.1016/J.MATPR.2019.12.355

[21] Huang Z, Sugiyama S, Yanagimoto J. Hybrid joining process for carbon fiber reinforced thermosetting plastic and metallic thin sheets by chemical bonding and plastic deformation. Journal of Materials Processing Technology. 2013;**213**:1864-1874. DOI:10.1016/J.JMATPROTEC.2013.04.015

[22] Ahrens M, Mallick V, Parfrey K. Robotic based thermoplastic fibre placement process. Proceedings of the IEEE International Conference Robotics Automation. 1998;**2**:1148-1153. DOI:10.1109/ROBOT.1998.677247

[23] Alauddin M, Choudhury IA, El Baradie MA, Hashmi MSJ. Plastics and their machining: A review. Journal of Materials Process Technology. 1995;**54**:40-46. DOI:10.1016/0924-0136(95)01917-0

[24] Greene JP. Commodity plastics. In: Automotive Plastics and Composites. 2021. pp. 83-105. DOI: 10.1016/B978-0-12-818008-2.00004-0

[25] Statista Global thermoplastic production by type. 2050. Available from: https://www.statista.com/statistics/1192886/thermoplastics-production-volume-by-type-globally/ [Accessed: October 09, 2022]

[26] Thermoplastics. Available from: https://www.bpf.co.uk/plastipedia/polymers/polymer-thermoplastics.aspx [Accessed: October 09, 2022]

[27] Plastics - the Facts 2022. In: Plast. Eur. Available from: https://plasticseurope.org/knowledge-hub/plastics-the-facts-2022/ [Accessed: January 31, 2023]

[28] Shi C, Zhang S, Zhao J, et al. Experimental study on removal of microplastics from aqueous solution by magnetic force effect on the magnetic sepiolite. Separation and Purification Technology. 2022;**288**:120564. DOI:10.1016/J.SEPPUR.2022.120564

[29] Ragusa A, Notarstefano V, Svelato A, et al. Raman microspectroscopy detection

and characterisation of microplastics in human breastmilk. Polymer. 2022;**14**:2700. DOI:10.3390/POLYM14132700

[30] Ben W, Zhu B, Yuan X, et al. Occurrence, removal and risk of organic micropollutants in wastewater treatment plants across China: Comparison of wastewater treatment processes. Water Research. 2018;**130**:38-46. DOI:10.1016/J.WATRES.2017.11.057

[31] Khanzada NK, Farid MU, Kharraz JA, et al. Removal of organic micropollutants using advanced membrane-based water and wastewater treatment: A review. Journal of Membrane Science. 2020;**598**:117672. DOI:10.1016/J.MEMSCI.2019.117672

[32] Tang W, Li H, Fei L, et al. The removal of microplastics from water by coagulation: A comprehensive review. Science Total Environment. 2022;**851**:158224. DOI:10.1016/J.SCITOTENV.2022.158224

[33] AlMaadeed MA, Ouederni M, Noorunnisa Khanam P. Effect of chain structure on the properties of glass fibre/polyethylene composites. Materials and Design. 2013;**47**:725-730. DOI:10.1016/J.MATDES.2012.11.063

[34] García E, Louvier-Hernández JF, Cervantes-Vallejo FJ, et al. Mechanical, dynamic and tribological characterization of HDPE/peanut shell composites. Polymer Testing. 2021;**98**:107075. DOI:10.1016/J.POLYMERTESTING.2021.107075

[35] Tazi M, Erchiqui F, Godard F, et al. Characterization of rheological and thermophysical properties of HDPE–wood composite. Journal of Applied Polymer Science. 2014;**131**:40495. DOI:10.1002/APP.40495

[36] Yu TH, Wilkes GL. Orientation determination and morphological study of high density polyethylene (HDPE) extruded tubular films: Effect of processing variables and molecular weight distribution. Polymer (Guildf). 1996;**37**:4675-4687. DOI:10.1016/S0032-3861(96)00307-2

[37] Pandharinath Said P, Chandra Pradhan R, Sharma N, Naik B. Protective coatings for shelf life extension of fruits and vegetables. Journal of Bioresources Engineering Technology. 2013;**1**:6

[38] Rasib SZM, Mariatti M, Atay HY. Effect of waste fillers addition on properties of high-density polyethylene composites: Mechanical properties, burning rate, and water absorption. Polymer Bulletin. 2021;**78**:6777-6795. DOI:10.1007/S00289-020-03454-3/FIGURES/10

[39] Adhikary KB, Pang S, Staiger MP. Dimensional stability and mechanical behaviour of wood-plastic composites based on recycled and virgin high-density polyethylene (HDPE). Composites. Part B, Engineering. 2008;**39**:807-815. DOI:10.1016/J.COMPOSITESB.2007.10.005

[40] Sahu AK, Sudhakar K, Sarviya RM. U.V Light Effect on the Mechanical Behaviour of HDPE/Carbon Black Composites. In: IOP Conference Series: Materials Science and Engineering. IOP Publishing; 2020. p. 012054. DOI: 10.1088/1757-899X/788/1/012054

[41] Zheng W, Lu X, Wong SC. Electrical and mechanical properties of expanded graphite-reinforced high-density polyethylene. Journal of Applied Polymer Science. 2004;**91**:2781-2788. DOI:10.1002/APP.13460

[42] Khouaja A, Koubaa A, Ben Daly H. Dielectric properties and thermal stability of cellulose high-density polyethylene bio-based composites. Industrial Crops and Products.

2021;**171**:113928. DOI:10.1016/J. INDCROP.2021.113928

[43] Lei Y, Wu Q, Clemons CM, et al. Influence of nanoclay on properties of HDPE/wood composites. Journal of Applied Polymer Science. 2007;**106**:3958-3966. DOI:10.1002/APP.27048

[44] Lim KLK, Mohd Ishak ZA, Ishiaku US, et al. High-density polyethylene/ultrahigh-molecular-weight polyethylene blend. I. the processing, thermal, and mechanical properties. Journal of Applied Polymer Science. 2005;**97**:413-425. DOI:10.1002/APP.21298

[45] Mohanty S, Verma SK, Nayak SK. Dynamic mechanical and thermal properties of MAPE treated jute/HDPE composites. Composites Science and Technology. 2006;**66**:538-547. DOI:10.1016/J. COMPSCITECH.2005.06.014

[46] Kusuktham B, Teeranachaideekul P. Mechanical properties of high density polyethylene/modified calcium silicate composites. SILICON. 2014;**6**:179-189. DOI:10.1007/S12633-014-9204-4

[47] Guo G, Finkenstadt VL, Nimmagadda Y. Mechanical properties and water absorption behavior of injection-molded wood fiber/carbon fiber high-density polyethylene hybrid composites. Advanced Composite Hybrid Materials. 2019;**2**:690-700. DOI:10.1007/S42114-019-00116-5/FIGURES/12

[48] Kanagaraj S, Varanda FR, Zhiltsova TV, et al. Mechanical properties of high density polyethylene/carbon nanotube composites. Composites Science and Technology. 2007;**67**:3071-3077. DOI:10.1016/j.compscitech.2007.04.024

[49] Batista NL, Helal E, Kurusu RS, et al. Mass-produced graphene—HDPE nanocomposites: Thermal, rheological, electrical, and mechanical properties. Polymer Engineering and Science. 2019;**59**:675-682. DOI:10.1002/PEN.24981

[50] Lange J, Wyser Y. Recent innovations in barrier Technologies for Plastic Packaging - a review. Packaging Technology and Science. 2003;**16**:149-158

[51] Wieme T, Duan L, Mys N, et al. Effect of matrix and graphite filler on thermal conductivity of industrially feasible injection molded thermoplastic composites. Polymers (Basel). 2019;**11**:87. DOI:10.3390/polym11010087

[52] Grigorescu RM, Grigore ME, Iancu L, et al. Waste electrical and electronic equipment: A review on the identification methods for polymeric materials. Recycling. 2019;**4**:32

[53] Mohan Bhasney S, Dhar P, Katiyar V. Polymer blends for sustainable food packaging. In: Sustainable Polymers for Food Packaging. De Gruyter; 2020. pp. 145-158. DOI: 10.1515/9783110648034-008

[54] Ziaabka M, Mertas A, Krol W, et al. High density polyethylene containing antibacterial silver nanoparticles for medical applications. Macromolecular Symposia. 2012;**315**:218-225. DOI:10.1002/MASY.201250527

[55] Murty MVS, Grulke EA, Bhattacharyya D. Influence of metallic additives on thermal degradation and liquefaction of high density polyethylene (HDPE). Polymer Degradation and Stability. 1998;**61**:421-430. DOI:10.1016/S0141-3910(97)00228-0

[56] Menderes A, Baytekin C, Topcu A, et al. Craniofacial reconstruction with high-density porous polyethylene implants. The Journal of Craniofacial Surgery. 2004;**15**:1336.

DOI:10.1097/00001665-200409000-00004

[57] Kimura K, Davis S, Thomas E, et al. 3D customization for Microtia repair in Hemifacial Microsomia. Laryngoscope. 2022;**132**:545-549. DOI:10.1002/LARY.29823

[58] Kumar NG, Sreenivas M, Gowda S. Cranioplasty of large cranial defects with porous polyethylene implants. The Journal of Craniofacial Surgery. 2016;**27**:e333-e335. DOI:10.1097/SCS.0000000000002480

[59] Villarreal PM, Monje F, Morillo AJ, et al. Porous polyethylene implants in orbital floor reconstruction. Plastic and Reconstructive Surgery. 2002;**109**:877-885; discussion 886. DOI:10.1097/00006534-200203000-00007

[60] Rubin PAD, Bilyk JR, Shore JW. Orbital reconstruction using porous polyethylene sheets. Ophthalmology. 1994;**101**:1697-1708. DOI:10.1016/S0161-6420(94)31113-4

[61] Romo T, Fozo MS, Sclafani AP. Microtia reconstruction using a porous polyethylene framework. Facial Plastic Surgery. 2000;**16**:15-22. DOI:10.1055/S-2000-7322/BIB

[62] Yaremchuk MJ. Facial skeletal reconstruction using porous polyethylene implants. Plastic and Reconstructive Surgery. 2003;**111**:1818-1827. DOI:10.1097/01.PRS.0000056866.80665.7A

[63] Niechajev I. Porous polyethylene implants for nasal reconstruction: Clinical and histologic studies. Aesthetic Plastic Surgery. 1999;**236**(23):395-402. DOI:10.1007/S002669900308

[64] Paxton NC, Allenby MC, Lewis PM, Woodruff MA. Biomedical applications of polyethylene. European Polymer Journal. 2019;**118**:412-428. DOI:10.1016/J.EURPOLYMJ.2019.05.037

[65] King Starboard HDPE Sheet Honolulu, Hawaii - Min Plastic Hawaii. Available from: https://minplastics.com/product/hdpe-starboard/ [Accessed: Octocer 20, 2022]

[66] High Density Polyethylene (HDPE) Foam Market to Witness Robust. Available from: https://www.openpr.com/news/1837661/high-density-polyethylene-hdpe-foam-market-to-witness-robust [Accessed: October 20, 2022]

[67] Fiber Optic Cable Protection Solutions (Microducts). esen.com. Available from: https://esen.com/en/products/fiber-optic-cable-protection-solutions-microducts/ [Accessed: October 21, 2022]

[68] High Density Polyethylene Pipes (PE). Available from: https://www.jains.com/Pipefittings/hdpe pipe.htm [Accessed: October 21, 2022]

[69] Better HDPE Bottles and Containers | TOPAS. Available from: https://topas.com/markets/packaging/better-hdpe-bottles-and-containers-0 [Accessed: Octover 24, 2022]

[70] Gupta N, Woldesenbet E, Mensah P. Compression properties of syntactic foams: Effect of cenosphere radius ratio and specimen aspect ratio. Composites. Part A, Applied Science and Manufacturing. 2004;**35**:103-111. DOI:10.1016/J.COMPOSITESA.2003.08.001

[71] Shutov FA. Syntactic polymer foams. Advances in Polymer Science. 1986;**1986**:63-123. DOI:10.1007/3-540-15786-7_7/COVER

[72] Bikiaris DN, Triantafyllidis KS. HDPE/Cu-nanofiber nanocomposites

with enhanced antibacterial and oxygen barrier properties appropriate for food packaging applications. Materials Letters. 2013;**93**:1-4. DOI:10.1016/J. MATLET.2012.10.128

[73] Jabłońska B, Kiełbasa P, Korenko M, Dróżdż T. Physical and chemical properties of waste from PET bottles washing as a component of solid fuels. Energies. 2019;**2197**(12):2197. DOI:10.3390/EN12112197

[74] General Chemical Resistance Chart | U.S. Plastic. Available from: https://www.usplastic.com/knowledgebase/article.aspx?contentkey=825 [Accessed: September 01, 2021]

[75] Chemical Resistance Chart | Plastics International. Available from: https://www.plasticsintl.com/chemical-resistance-chart [Accessed: September 01, 2021]

[76] Chemical Resistance Chart for Plastics | Chemical Compatibility. Available from: https://www.curbellplastics.com/Research-Solutions/Chemical-Resistance-of-Plastics [Accessed: September 01, 2021]

[77] Colwill J, Simeone A, Gould O, et al. Energy-efficient Systems for the Sensing and Separation of mixed polymers. Procedia CIRP. 2017;**62**:512-517. DOI:10.1016/J.PROCIR.2016.06.116

[78] Sudakov A, Chudyk I, Sudakova D, Dziubyk L. Innovative technology for insulating the borehole absorbing horizons with thermoplastic materials. E3S Web Conference. 2019;**123**:1-13. DOI:10.1051/E3SCONF/201912301033

[79] Yusof F, Miyashita Y, Seo N, et al. Utilising friction spot joining for dissimilar joint between aluminium alloy (A5052) and polyethylene terephthalate. Science and Technology of Welding and Joining. 2012;**17**:544-549. DOI:10.1179/136217112X13408696326530

[80] Chowreddy RR, Nord-Varhaug K, Rapp F. Recycled poly(ethylene terephthalate)/clay nanocomposites: Rheology, thermal and mechanical properties. Journal of Polymers and the Environment. 2018;**271**(27):37-49. DOI:10.1007/S10924-018-1320-6

[81] Awaja F, Pavel D. Recycling of PET. European Polymer Journal. 2005;**41**:1453-1477. DOI:10.1016/J. EURPOLYMJ.2005.02.005

[82] Hsiao K-J, Jen Z-F, Lu C-L. Analysis of physical properties of PET/CD-PET polyblend hollow fiber and its kinetics of alkaline hydrolysis. Journal of Applied Polymer Science. 2002;**86**:3601-3610. DOI:10.1002/APP.11226

[83] Olam M, Karaca H. Effect of sodium boron hydride (NaBH4) on waste polyethylene terephthalate pyrolysis. IOP Conference Series Earth Environmental Science. 2019;**362**:012032. DOI:10.1088/1755-1315/362/1/012032

[84] Oh D, Lee HW, Kim YM, Park YK. Catalytic Pyrolysis of Polystyrene and Polyethylene Terephthalate over Al-MSU-F. Elsevier Ltd; 2018. pp. 111-117. DOI: 10.1016/j.egypro.2018.06.015

[85] Kannan P, Al Shoaibi A, Srinivasakannan C. Energy recovery from co-gasification of waste polyethylene and polyethylene terephthalate blends. Computers and Fluids. 2013;**88**:38-42. DOI:10.1016/j. compfluid.2013.09.004

[86] Yang T, Hu K, Li R, et al. Cogasification of typical plastics and rice straw with carbon dioxide. Environmental Progress & Sustainable Energy. 2015;**34**:789-794. DOI:10.1002/EP.12041

[87] Izzatie NI, Basha MH, Uemura Y, et al. Co-pyrolysis of rice straw and polyethylene terephthalate (PET) using a fixed bed drop type pyrolyzer. Journal of Physics Conference Series. 2017;**908**:012073. DOI:10.1088/1742-6596/908/1/012073

[88] Velásquez E, Garrido L, Valenzuela X, et al. Physical properties and safety of 100% post-consumer PET bottle -organoclay nanocomposites towards a circular economy. Sustainable Chemistry and Pharmacy. 2020;**17**:100285. DOI:10.1016/J.SCP.2020.100285

[89] Nakajima H, Dijkstra P, Loos K. The recent developments in biobased polymers toward general and engineering applications: Polymers that are upgraded from biodegradable polymers, analogous to petroleum-derived polymers, and newly developed. Polymers (Basel). 2017;**9**:1-26. DOI:10.3390/POLYM9100523

[90] PET polyester plastic - TECAPET | Ensinger. Available from: https://www.ensingerplastics.com/en/shapes/engineering-plastics/pet-polyester [Accessed: September 02, 2021]

[91] Saidi MAA, Hassan A, Wahit MU, et al. Thermal, dynamic mechanical analysis and mechanical properties of polybutylene terephthalate/polyethylene terephthalate blends. Journal of Teknology. 2020;**82**:73-83. DOI:10.11113/JT.V82.14802

[92] Thongsong W, Kulsetthanchalee C, Threepopnatkul P. Effect of polybutylene adipate-co-terephthalate on properties of polyethylene terephthalate thin films. Materials Today Proceedings. 2017;**4**:6597-6604. DOI:10.1016/J.MATPR.2017.06.173

[93] Gorrasi G, Bugatti V, Milone C, et al. Effect of temperature and morphology on the electrical properties of PET/conductive nanofillers composites. Composites. Part B, Engineering. 2018;**135**:149-154. DOI:10.1016/j.compositesb.2017.10.020

[94] Liu RYF, Hu YS, Hibbs MR, et al. Improving oxygen barrier properties of poly(ethylene terephthalate) by incorporating isophthalate. I. Effect of orientation. Journal of Applied Polymer Science. 2005;**98**:1615-1628. DOI:10.1002/APP.22213

[95] Gorrasi G, Milone C, Piperopoulos E, Pantani R. Preparation, processing and analysis of physical properties of calcium ferrite-CNTs/PET nano-composite. Composites. Part B, Engineering. 2015;**81**:44-52. DOI:10.1016/J.COMPOSITESB.2015.06.019

[96] Paszkiewicz S, Nachman M, Szymczyk A, et al. Influence of expanded graphite (EG) and graphene oxide (GO) on physical properties of PET based nanocomposites. Polish Journal of Chemical Technology. 2014;**16**:45-50. DOI:10.2478/PJCT-2014-0068

[97] Michiels Y, Van Puyvelde P, Sels B. Barriers and chemistry in a bottle: Mechanisms in today's oxygen barriers for tomorrow's materials. Applied Sciences. 2017;**7**:665

[98] Xing L, Wang Y, Wang S, et al. Effects of modified graphene oxide on thermal and crystallization properties of PET. Polymer. 2018;**10**:613. DOI:10.3390/POLYM10060613

[99] Wang Y, Wang W, Zhang Z, et al. Study of the glass transition temperature and the mechanical properties of PET/modified silica nanocomposite by molecular dynamics simulation. European Polymer Journal. 2016;**75**:36-45. DOI:10.1016/j.eurpolymj.2015.11.038

[100] Chen H, Pyda M, Cebe P. Non-isothermal crystallization of PET/PLA blends. Thermochimica Acta. 2009;**492**:61-66. DOI:10.1016/j.tca.2009.04.023

[101] Kilinc M, Cakal GO, Bayram G, et al. Flame retardancy and mechanical properties of pet-based composites containing phosphorus and boron-based additives. Journal of Applied Polymer Science. 2015;**132**:42016. DOI:10.1002/APP.42016

[102] Jogur G, Nawaz Khan A, Das A, et al. Impact properties of thermoplastic composites. Textile Progress. 2018;**50**:109-183. DOI:10.1080/00405167.2018.1563369

[103] Misri Z, Ibrahim MHW, Halid AA, et al. Dynamic mechanical analysis of waste polyethylene terephthalate bottle. International Journal of Integrated Engineering. 2018;**10**:38-42. DOI:10.30880/IJIE.2018.10.09.023

[104] Nayak S, Jena PK, Samal P, et al 2021. Improvement of Mechanical and Thermal Properties of Polyethylene Terephthalate (PET) Composite Reinforced with Chemically Treated Ladies Finger Natural Fiber. 101080/15440478202119332680

[105] Huq T, Khan A, Akter T, et al. Thermo-mechanical, degradation, and interfacial properties of jute fiber-reinforced PET-based composite. Journal of Thermoplastic Composite Materials. 2011;**24**:889-898. DOI:10.1177/0892705711401846

[106] Wang LN, Xin CZ, Liu WT, et al. (2015) electrospun PET/PEG fibrous membrane with enhanced mechanical properties and hydrophilicity for filtration applications. Arabian Journal for Science and Engineering. 2015;**4010**(40):2889-2895. DOI:10.1007/S13369-015-1828-1

[107] PET (Thermoplastic Polyester) | Poly-Tech Industrial. Available from: https://www.polytechindustrial.com/products/plastic-stock-shapes/pet-thermoplastic-polyester [Accessed: September 06, 2021]

[108] Chung S-C, Hahm W-G, Im S-S, Oh S-G. Poly(ethylene terephthalate) (PET) nanocomposites filled with fumed silicas by melt compounding. Macromolecular Research. 2002;**104**(10):221-229. DOI:10.1007/BF03218309

[109] Polyethylene terephthalate - online catalogue source - supplier of research materials in small quantities - Goodfellow. Available from: http://www.goodfellow.com/E/Polyethylene-terephthalate.html [Accessed: August 19, 2021]

[110] Lima GR, Monteiro WF, Toledo BO, et al. Titanate nanotubes modified with zinc and its application in post-consumer PET depolymerization. Macromolecular Symposia. 2019;**383**:1800008. DOI:10.1002/MASY.201800008

[111] Stocco A, La CV, Piccarolo S, Brucato V. The solidification behavior of a PBT/PET blend over a wide range of cooling rate. Journal of Polymer Science. 2009;**47**:799-810. DOI:10.1002/POLB.21687

[112] Viscoelastic Materials. Available from: https://polymerdatabase.com/polymer physics/Modulus.html [Accessed: September 06, 2021]

[113] Kalia S, Fu SY. Cryogenic processing: State of the art, advantages and applications. In: Polymers at Cryogenic Temperatures. Berlin Heidelberg, Berlin, Heidelberg: Springer; 2012. pp. 1-7

[114] Kawai F, Kawabata T, Oda M. Current knowledge on enzymatic PET degradation and its possible application

to waste stream management and other fields. Applied Microbiology and Biotechnology. 2019;**103**:4253-4268. DOI:10.1007/S00253-019-09717-Y/FIGURES/1

[115] Ghosal K, Nayak C. Recent advances in chemical recycling of polyethylene terephthalate waste into value added products for sustainable coating solutions – hope vs . hype. Materials Advance. 2022;**3**:1974-1992. DOI:10.1039/D1MA01112J

[116] Sang T, Wallis CJ, Hill G, Britovsek GJP. Polyethylene terephthalate degradation under natural and accelerated weathering conditions. European Polymer Journal. 2020;**136**:109873. DOI:10.1016/J.EURPOLYMJ.2020.109873

[117] Kim MO, Park JK, Han TH, et al. Influence of polyethylene terephthalate powder on hydration of Portland cement. Polymers (Basel). 2021;**13**. DOI:10.3390/POLYM13152551

[118] Hadjizadeh A, Ajji A, Bureau MN. Nano/micro electro-spun polyethylene terephthalate fibrous mat preparation and characterization. Journal of the Mechanical Behavior of Biomedical Materials. 2011;**4**:340-351. DOI:10.1016/J.JMBBM.2010.10.014

[119] Ng R, Zhang X, Liu N, Yang S-T Modifications of nonwoven polyethylene terephthalate fibrous matrices via NaOH hydrolysis: Effects on pore size, fiber diameter, cell seeding and proliferation. 10.1016/j.procbio.2009.04.024

[120] Lee SJ, Yoo JJ, Lim GJ, et al. In vitro evaluation of electrospun nanofiber scaffolds for vascular graft application. Journal of Biomedical Material Research - Part A. 2007;**83**:999-1008. DOI:10.1002/JBM.A.31287

[121] Andrews KD, Feugier P, Black RA, Hunt JA. Vascular prostheses: Performance related to cell-shear responses. The Journal of Surgical Research. 2008;**149**:39-46. DOI:10.1016/J.JSS.2007.08.030

[122] Aortic vascular prosthesis - AlboGraft® - LeMaitre Vascular - synthetic / birfurcated. Available from: https://www.medicalexpo.com/prod/lemaitre-vascular/product-78766-487338.html [Accessed: October 14, 2022]

[123] Arun J, Gopinath KP, SundarRajan PS, et al. Co-liquefaction of Prosopis juliflora with polyolefin waste for production of high grade liquid hydrocarbons. Bioresource Technology. 2019;**274**:296-301. DOI:10.1016/J.BIORTECH.2018.11.102

[124] PET Chips Export Harming Carpet Industry | Financial Tribune. Available from: https://financialtribune.com/articles/economy-business-and-markets/72004/pet-chips-export-harming-carpet-industry [Accessed: October 14, 2022]

[125] PET for Water Bottles - Jade PET resin China. Available from: http://jadepetresin.com/pet-for-water-bottles/ [Accessed: October 24, 2022]

[126] Serda M, Becker FG, Cleary M, et al. Food Packaging Materials | Point Five Packaging. LLC, Food Packaging Materials; 2013. Available from: https://www.p5pkg.com/materials/ [Accessed: October 24, 2022]

[127] Hadjizadet A, Doillon CJ, Vermette P. Bioactive polymer fibers to direct endothelial cell growth in a three-dimensional environment. Biomacromolecules. 2007;**8**:864-873. DOI:10.1021/BM060957F/ASSET/IMAGES/MEDIUM/BM060957FN00001.GIF

[128] Çaykara T, Sande MG, Azoia N, et al. Exploring the potential of

polyethylene terephthalate in the design of antibacterial surfaces. Medical Microbiology and Immunology. 2020;**209**:363. DOI:10.1007/S00430-020-00660-8

[129] Castelvetro V, Corti A, Bianchi S, et al. Quantification of poly(ethylene terephthalate) micro- and nanoparticle contaminants in marine sediments and other environmental matrices. Journal of Hazardous Materials. 2020;**385**:121517. DOI:10.1016/J.JHAZMAT.2019.121517

[130] Man-Made Fibers Continue To Grow | Textile World. Available from: https://www.textileworld.com/textile-world/fiber-world/2015/02/man-made-fibers-continue-to-grow/ [Accessed: October 14, 2022]

[131] Smithers. The Future of PET Packaging to 2025 | Market Reports and Trends. 2020. Available from: https://www.smithers.com/resources/2020/sept/global-pet-packaging-demand-to-reach-$44-1-billion [Accessed: September 13, 2021]

[132] Yuan Z, Nag R, Cummins E. Ranking of potential hazards from microplastics polymers in the marine environment. Journal of Hazardous Materials. 2022;**429**:128399. DOI:10.1016/J.JHAZMAT.2022.128399

[133] Bajt O. From plastics to microplastics and organisms. FEBS Open Bio. 2021;**11**:954. DOI:10.1002/2211-5463.13120

[134] Khalid N, Aqeel M, Noman A, et al. Interactions and effects of microplastics with heavy metals in aquatic and terrestrial environments. Environmental Pollution. 2021;**290**. DOI:10.1016/J.ENVPOL.2021.118104

[135] Browne MA, Galloway T, Thompson R. Microplastic—An emerging contaminant of potential concern? Integrated Environmental Assessment and Management. 2007;**3**:559-561. DOI:10.1002/IEAM.5630030412

[136] Singh B, Sharma N. Mechanistic implications of plastic degradation. Polymer Degradation and Stability. 2008;**93**:561-584. DOI:10.1016/J.POLYMDEGRADSTAB. 2007.11.008

[137] Andrady AL. Microplastics in the marine environment. Marine Pollution Bulletin. 2011;**62**:1596-1605. DOI:10.1016/J.MARPOLBUL.2011.05.030

[138] Barnes DKA, Galgani F, Thompson RC, Barlaz M. Accumulation and fragmentation of plastic debris in global environments. Philosophical Transaction on Royal Society B Biological Science. 2009;**364**:1985-1998. DOI:10.1098/RSTB.2008.0205

[139] Grigoriadou I, Pavlidou E, Paraskevopoulos KM, et al. Comparative study of the photochemical stability of HDPE/Ag composites. Polymer Degradation and Stability. 2018;**153**:23-36. DOI:10.1016/J.POLYMDEGRADSTAB. 2018.04.016

[140] Ainali NM, Bikiaris DN, Lambropoulou DA. Aging effects on low- and high-density polyethylene, polypropylene and polystyrene under UV irradiation: An insight into decomposition mechanism by Py-GC/MS for microplastic analysis. Journal of Analytical and Applied Pyrolysis. 2021;**158**:105207. DOI:10.1016/J.JAAP.2021.105207

[141] Brandon J, Goldstein M, Ohman MD. Long-term aging and degradation of microplastic particles: Comparing in situ oceanic and experimental weathering patterns. Marine Pollution Bulletin.

2016;**110**:299-308. DOI:10.1016/J.MARPOLBUL.2016.06.048

[142] Grigoriadou I, Paraskevopoulos KM, Chrissafis K, et al. Effect of different nanoparticles on HDPE UV stability. Polymer Degradation and Stability. 2011;**96**:151-163. DOI:10.1016/J.POLYMDEGRADSTAB. 2010.10.001

[143] Fabiyi JS, McDonald AG, Wolcott MP, Griffiths PR. Wood plastic composites weathering: Visual appearance and chemical changes. Polymer Degradation and Stability. 2008;**93**:1405-1414. DOI:10.1016/J.POLYMDEGRADSTAB. 2008.05.024

[144] Stark NM, Matuana LM. Surface chemistry changes of weathered HDPE/wood-flour composites studied by XPS and FTIR spectroscopy. Polymer Degradation and Stability. 2004;**86**:1-9. DOI:10.1016/J.POLYMDEGRADSTAB. 2003.11.002

[145] Hu X. Wavelength sensitivity of photo-oxidation of polyethylene. Polymer Degradation and Stability. 1997;**55**:131-134. DOI:10.1016/S0141-3910(96)00120-6

[146] Chelliah A, Subramaniam M, Gupta R, Gupta A. Evaluation on the Thermo-oxidative degradation of PET using Prodegradant additives. Indian Journal of Science and Technology. 2017;**10**:1-5. DOI:10.17485/IJST/2017/V10I6/111212

[147] Fukushima K, Tabuani D, Dottori M, et al. Effect of temperature and nanoparticle type on hydrolytic degradation of poly(lactic acid) nanocomposites. Polymer Degradation and Stability. 2011;**96**:2120-2129. DOI:10.1016/J.POLYMDEGRADSTAB. 2011.09.018

[148] Balasubramanian V, Natarajan K, Rajeshkannan V, Perumal P. Enhancement of in vitro high-density polyethylene (HDPE) degradation by physical, chemical, and biological treatments. Environmental Science and Pollution Research. 2014;**21**:12549-12562. DOI:10.1007/S11356-014-3191-2/FIGURES/8

[149] Ojha N, Pradhan N, Singh S, et al. (2017) evaluation of HDPE and LDPE degradation by fungus, implemented by statistical optimization. Scientific Reports. 2017;**71**(7):1-13. DOI:10.1038/srep39515

[150] Moog D, Schmitt J, Senger J, et al. Using a marine microalga as a chassis for polyethylene terephthalate (PET) degradation. Microbial Cell Factories. 2019;**18**:1-15. DOI:10.1186/S12934-019-1220-Z/FIGURES/5

[151] Shabbir S, Faheem M, Ali N, et al. Periphytic biofilm: An innovative approach for biodegradation of microplastics. Science Total Environment. 2020;**717**:137064. DOI:10.1016/J.SCITOTENV. 2020.137064

[152] Farzi A, Dehnad A, Fotouhi AF. Biodegradation of polyethylene terephthalate waste using Streptomyces species and kinetic modeling of the process. Biocatalysis and Agricultural Biotechnology. 2019;**17**:25-31. DOI:10.1016/J.BCAB.2018.11.002

[153] Farzi A, Dehnad A, Shirzad N, Norouzifard F. Biodegradation of high density polyethylene using Streptomyces species. Journal of Coast Life Medicine. 2017:474-479. DOI:10.12980/JCLM.5.2017J7-94

[154] Lin TR, Lin TA, Lin MC, et al. Impact resistance of fiber reinforced sandwich-structured nonwoven composites: Reinforcing effect of different fiber length. Materials Today

Communication. 2020;**24**:101345. DOI:10.1016/J.MTCOMM.2020.101345

[155] Jaiswal S, Sharma B, Shukla P. Integrated approaches in microbial degradation of plastics. Environmental Technology and Innovation. 2020;**17**:100567. DOI:10.1016/J.ETI.2019.100567

[156] Pathak SJ, Gangal A, Prabu V. Direct liquefaction of discarded printer cartridge plastics and its kinetic modelling. Fuel Processing Technology. 2022;**228**:107147. DOI:10.1016/J.FUPROC.2021.107147

[157] Olam M, Karaca H. Optimization of process parameters at direct liquefaction of waste PETs. Process Safety and Environment Protection. 2022;**2022**. DOI:10.1016/J.PSEP.2022.10.058

[158] Chen Y, Wen D, Pei J, et al. Identification and quantification of microplastics using Fourier-transform infrared spectroscopy: Current status and future prospects. Current Opinion in Environmental Science Health. 2020;**18**:14-19. DOI:10.1016/J.COESH.2020.05.004

[159] Zhang C, Chen X, Wang J, Tan L. Toxic effects of microplastic on marine microalgae Skeletonema costatum: Interactions between microplastic and algae. Environmental Pollution. 2017;**220**:1282-1288. DOI:10.1016/J.ENVPOL.2016.11.005

[160] (PDF) Microwave Energy Applied to Processing of High-Temperature Materials. Available from: https://www.researchgate.net/publication/265280477_Microwave_Energy_Applied_to_Processing_of_High-Temperature_Materials [Accessed: September 10, 2021]

[161] Gasperi J, Wright SL, Dris R, et al. Microplastics in air: Are we breathing it in? Current Opinion in Environment Science Health. 2018;**1**:1-5. DOI:10.1016/J.COESH.2017.10.002

[162] Sharma MD, Elanjickal AI, Mankar JS, Krupadam RJ. Assessment of cancer risk of microplastics enriched with polycyclic aromatic hydrocarbons. Journal of Hazardous Materials. 2020;**398**:122994. DOI:10.1016/J.JHAZMAT.2020.122994

[163] Khalid N, Aqeel M, Noman A, et al. Linking effects of microplastics to ecological impacts in marine environments. Chemosphere. 2021;**264**:128541. DOI:10.1016/J.CHEMOSPHERE.2020.128541

[164] Lozano YM, Rillig MC. Effects of microplastic Fibers and drought on plant communities. Environmental Science & Technology. 2020;**54**:6166-6173. DOI:10.1021/ACS.EST.0C01051/ASSET/IMAGES/LARGE/ES0C01051_0004.JPEG

[165] Issac MN, Kandasubramanian B. Effect of microplastics in water and aquatic systems. Environmental Science and Pollution Research. 2021, 2021;**2816**(28):19544-19562. DOI:10.1007/S11356-021-13184-2

[166] Cheng Y, Song W, Tian H, et al. The effects of high-density polyethylene and polypropylene microplastics on the soil and earthworm Metaphire guillelmi gut microbiota. Chemosphere. 2021;**267**:129219. DOI:10.1016/J.CHEMOSPHERE.2020.129219

[167] Bringer A, Thomas H, Prunier G, et al. High density polyethylene (HDPE) microplastics impair development and swimming activity of Pacific oyster D-larvae, Crassostrea gigas, depending on particle size. Environmental Pollution. 2020;**260**:113978. DOI:10.1016/J.ENVPOL.2020.113978

[168] Kim Y, Jeong J, Lee S, et al. Identification of adverse outcome pathway related to high-density polyethylene microplastics exposure: Caenorhabditis elegans transcription factor RNAi screening and zebrafish study. Journal of Hazardous Materials. 2020;**388**:121725. DOI:10.1016/J.JHAZMAT.2019.121725

[169] Jemec A, Horvat P, Kunej U, et al. Uptake and effects of microplastic textile fibers on freshwater crustacean Daphnia magna. Environmental Pollution. 2016;**219**:201-209. DOI:10.1016/J.ENVPOL.2016.10.037

[170] Shen J, Liang B, Zhang D, et al. Effects of PET microplastics on the physiology of Drosophila. Chemosphere. 2021;**283**:131289. DOI:10.1016/J.CHEMOSPHERE.2021.131289

[171] Weber A, Scherer C, Brennholt N, et al. PET microplastics do not negatively affect the survival, development, metabolism and feeding activity of the freshwater invertebrate Gammarus pulex. Environmental Pollution. 2018;**234**:181-189. DOI:10.1016/J.ENVPOL.2017.11.014

Chapter 6

The Risks of Microplastic Pollution in the Aquatic Ecosystem

Paul Agbekpornu and Isaac Kevudo

Abstract

Microplastic pollution is a global issue that has a detrimental effect on the food chain in the marine ecosystem. They are found in their highest concentrations along coastal lines and within mid-ocean gyres. In marine environments, microplastics are a threat to marine organisms, as they are often in the same size range as prey and are mistaken as food. When ingested can have a deleterious range of effects on marine organisms, a process which may facilitate the transfer of chemical additives or hydrophobic waterborne pollutants to aquatic lives. In this chapter, we looked at the risk of microplastic pollution and its impact on marine organisms and humankind. The study shows that consumption of microplastics has led to ingestion of chemical toxins in aquatic fish, which leads to damage of digestive organs, choking of marine organisms, channel for the spread of microbes, and a reduction in growth and reproductive output. These threats increase the risk to aquatic fishes and human survival. Hence, the need to educate the public on the dangers of using products that pose an immediate and long-term threat to the marine ecosystem and the health of its organism, and the food we eat by marine scientists.

Keywords: microplastic pollution, marine ecosystem, food chain, aquatic, microbeads, cosmetics

1. Introduction

Plastics are synthetic polymers that are pliable (flexible) in nature and may be molded into various shapes [1]. Plastic is made up of long chains of polymers made up of carbon, oxygen, hydrogen, silicon, and chloride, which are derived from natural gas, oil, and coal [2]. Polyethylene (PE), polyamide (PA), polypropylene (PP), polyester (PES) polyurethane (PU), acrylic (AC), polystyrene (PS), polyethylene terephthalate (PET), polyvinyl chloride (PVC), high-density polyethylene (HDPE), low-density polyethylene (LDPE), polyimide (PI), poly-methyl methacrylate (PMMA), polytetrafluoroethylene (PFE), polyvinyl chloride (PVC), and high-density polyethylene (HDPE) are the most common synthetic polymers, accounting for 90% of global plastic manufacturing [3, 4]. Materials made from plastics are cheap. And because of its low production cost, simplicity, durability, strength, corrosion resistance, good thermal and electrical conductivity, and physiochemical properties, plastic has become an essential and general material in all aspects of our lives.

The mass production of plastics began in the 1940s, and thereafter microplastic pollution of the marine ecosystem has been a growing problem [5]. Global plastic production has nearly tripled in the last three decades, and it is expected to reach 33 billion tons by 2050 [6–8]. Despite rising knowledge of plastic pollution and measures to reduce it, annual plastic output continues to rise. Research conducted by [4] has revealed that over 280 million metric tons of plastics wastes are generated by the manufacturing industrial sector yearly. An estimated amount of 275 million metric tons of territory plastic garbage from 192 coastal countries entered the marine, resulting in 4.8–12.7 million metric tons [9]. The percentage of worldwide aquatic plastic pollutants entering into the marine ecosystem [10] based on data published recently has been shown in **Figure 1**. The majority of the world's largest polluting water bodies are in Asia, with a few in Africa as well.

Microplastics are microscopic plastic pieces with diameters of 5 mm found in marine environments [11]. These microscopic plastics can be ingested by a variety of marine living organisms, including corals, planktons, marine invertebrates, fish, and whales, and are then passed through the food chain. These biodegradable plastics directly endanger marine species and have an indirect influence on the ecosystem by decontaminating other marine pollutants. Microplastics accumulate hydrophobic contaminants from the aquatic environment due to their huge surface area-to-volume ratio [12]. Thus, microplastic contamination is becoming a source of concern due to its negative impact, particularly on marine life.

2. Microplastics

Microplastics are pieces of plastic that are between a millimeter and a nanometer in size and are invisible to the human eye. The term "microplastics" has been defined differently by various researchers (see [13–16]), including a workshop on the topic. Microplastics are defined as being in the size range <5 mm [17] (recognizing 333

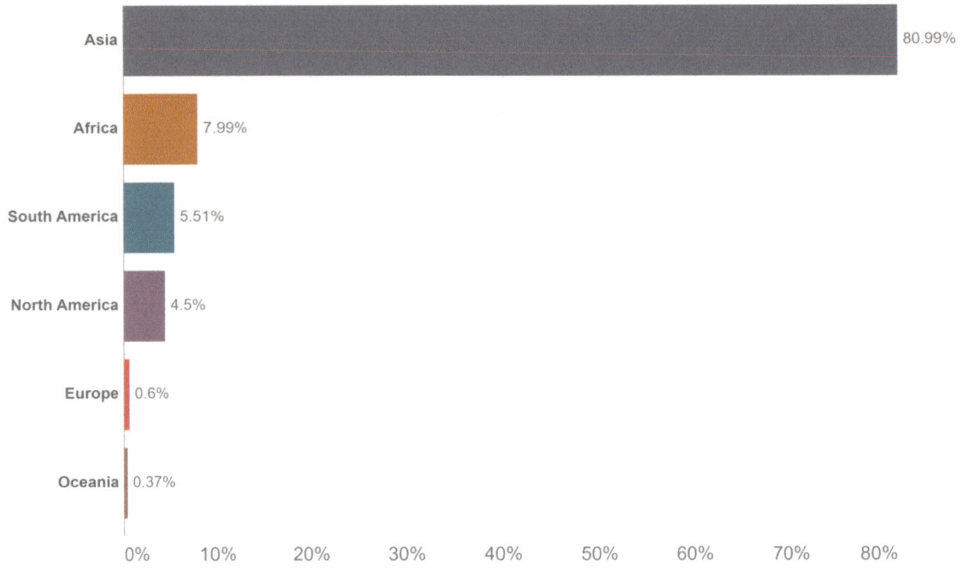

Figure 1.
Over 1000 rivers are accountable for 80% of worldwide aquatic plastic pollution in the ocean.

nanometers as a practical lower limit when neuston nets are used for sampling). Particles of plastics of sizes ranging from a few nanometers to 500 nanometers (5 mm) are commonly present in marine waters [18, 19]. For better understanding, the size range stated above is referred to as "microplastics" here. Other larger particles such as virgin resin pellets are referred to as "mesoplastics" [14].

Also, microplastics are tiny plastic granules used as scrubbers in cosmetics and air-blasting, and small plastic fragments are derived from the breakdown of macroplastics [20–22]. The presence of small plastic fragments in the open ocean was first highlighted in the 1970s [23], and a renewed scientific interest in microplastics over the past decade has revealed that these contaminants are widespread and ubiquitous within the marine environment, with the potential to cause harm to aquatic lives [24, 25]. Typically, these are the smaller pieces of bigger plastic objects, which are introduced into the marine ecosystem by a variety of mechanisms, such as industrial processes, human clothes (microfibers), and cosmetics (small beaded plastic). Due to their microscopic nature, they gradually make their way through the water systems where they are not cleaned out before being pumped back into the drainage channels. Persistent organic pollutants (POPs) that occur globally in marine waters at very low concentrations are picked up by these meso-microplastics via partitioning. The hydrophobicity of POPs facilitates their concentration in the meso-microplastic litters at a level higher than that in the marine ecosystem [26]. These contaminated plastics when ingested by aquatic organisms pose a serious problem by which the POPs can enter the marine food web. Unlike macroplastics, microplastics are not readily visible to the naked eye; even resin-pellets (mesoplastics) mixed with sand are not easily discernible. Net sampling does not of course collect the smaller microplastics and no acceptable standard procedure is presently available for their enumeration in water or sand [26].

2.1 Types of microplastics

Microplastics are classified into two groups based on their origin: primary and secondary microplastics [27].

2.1.1 Primary microplastics

Primary microplastics are micro-sized synthetic polymers that are directly introduced into the environment as minute particles. They are utilized as exfoliates in a variety of operations, such as chemical compositions, abrasive media, chemical and petrochemical cleaning, and synthetic clothing manufacturing. They can be added voluntarily to items like cleaning agents in hygiene and cosmetics (e.g., shower gels). They can also be caused by the abrasion of big plastic objects during manufacturing, usage, or maintenance, such as tyre erosion, while driving or the abrasion of synthetic textiles during washing [28]. Microbeads are a form of primary plastic (size 2 mm) that is made up of polyethylene (PE), polypropylene (PP), and polystyrene (PS) beads and are used in cosmetic and health care goods [12].

2.1.2 Secondary microplastics

Secondary microplastics are microplastics that result from the decomposition of larger plastic products into microscopic plastic pieces in the marine ecosystem. This occurs as a result of ecological changes, such as microbial degradation, photocatalysis,

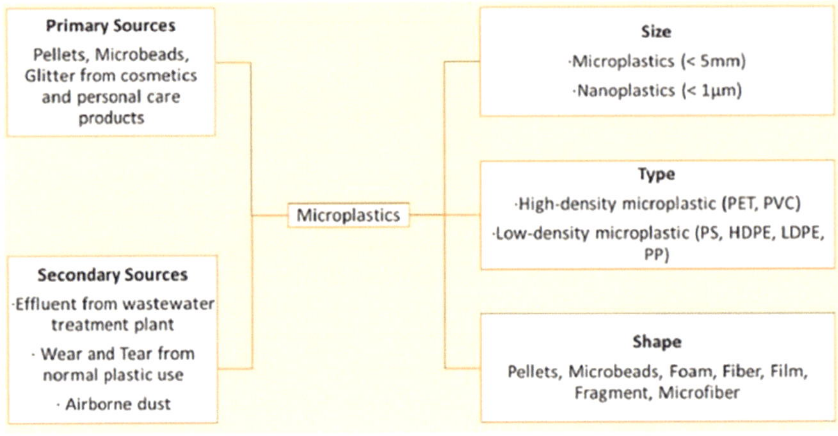

Figure 2.
Classification of microplastic based on sources, size, type, and shape [29].

high-temperature degradation, thermal decomposition, hydrolysis, and other weathering processes of indiscriminate dumping, such as abandoned plastic bags or unexpected losses, such as fishing gear. Microplastics in the waters can either circulate or sink. Microplastics that are lighter than seawater, such as polypropylene, will flow and spread across the oceans. They subsequently congregate in gyres formed by tidal currents (**Figure 2**) [28].

3. Sources of microplastics emission into the marine ecosystem

General littering, plastic waste mismanagement, tires, synthetic textiles, marine coatings, road markings, personal care products, plastic pellets, city dust, and release of wastewater from sewage treatment plants have been the main sources of microplastic pollution in the marine ecosystem [30]. Marine litter results from the indiscriminate disposal of refuses that are either directly or indirectly transferred to our seas and oceans [22, 31]. Whilst the emphasis of this study is on microplastics, in this section, we also consider the indiscriminate disposal of macroplastics, as with time, they eventually degrade into mesoplastics and microplastics. Plastic litter from land [terrestrial] sources contributes 80% of the plastics found in marine litter [26]. Such plastics include primary microplastics used in cosmetics and air-blasting, improperly disposed "user" plastics, and plastic leachates from refuse sites. With approximately half the world's population residing within 50 miles away from the coast, these kinds of plastic have a high likelihood of entering the marine ecosystem via rivers and wastewater systems, or by being blown off-shore [16, 32]. Plastic microbeads are utilized as components in cosmetic and personal care products for a range of functions, including an adsorptive state for active substance distribution, exfoliating, and viscosity control. Some products carry quite so much plastic as ingredients in which they are packaged [33]. These account for up to 10% of the product weight and thousands of microbeads per gram of product [34]. Microplastics used both in these cosmetics and as air-blasting media can easily enter waterways through domestic or industrial drainage systems [21]. The traditional use of products for

Figure 3.
Microplastic and garbage pollution present in the ocean [35].

personal care culminates in the direct input of microplastic into industrial wastewater from homes, hotels, hospitals, and sports facilities, such as beaches. Cosmetic microbeads have been detected in field investigations conducted in many parts of the world (**Figure 3**) [36].

Tourism and recreational activities had contributed to the discarded plastics left along beaches and coastal resorts [21], as well as those from marine debris, observed on beaches arising from the beaching of materials carried on in-shore and ocean currents [32]. Whilst wastewater treatment plants will trap macroplastics and some small plastic debris within sewage sludge, a greater percentage of microplastics will pass through such filtration systems [13, 37, 38]. Plastics that enter river systems—either directly or indirectly—will then be transported out into the ocean. A couple of studies conducted have shown how the high single-directional flow of freshwater systems drives the movement of plastic debris into the oceans [39, 40]. Another common marine source of plastic debris is fishing gear [26]. Discarded or lost fishing gear, including plastic monofilament line and nylon netting, which is typically buoyant and can therefore drift at variable depths within the oceans. This is particularly problematic due to its inherent capacity for causing entanglement of marine organisms, known as "ghost fishing" [31]. Historically, marine materials have been a great contributor to marine litter, with estimates indicating that during the 1970s the global commercial fishing fleet dumped over 23,000 tons of plastic packaging materials in the ocean [41]. Additionally, the manufacture of plastic products that use granules and small resin pellets, known as "nibs," as their raw material is another source of plastic debris [microplastics] [41–43]. Many plastics are introduced into the marine as pellets (usually 2–5 mm in diameter) or powders. Pellets are discharged into the marine environment through little or big occurrences along the entire plastic value chain during manufacturing, processing, transport, and recycling [44]. In the US alone, production rose from 2.9 million pellets in 1960 to 21.7 million pellets by 1987 [41].

4. Distribution of microplastics in the marine environment

Plastic contamination and microplastics have spread throughout the world's aquatic ecosystems [21, 22, 31]. Plastic pollution and microplastics can be transported over long distances by ocean currents, winds, river outflow, and drift [19, 45, 46],

Shape of microplastic	Type of marine ecosystem	Type of microplastic	Source of transport	References
microfiber	Deep sea Atlantic ocean	PS, PA, AC, acetate	Sewage treatment plants	[48]
Microplastic fragments	Tamar estuary	PVC, PES, PA	Wind, wave, and tides	[49, 50]
Microfibers	Shorelines	PES, AC fibers	Sewage treatment plants	[50]
Fibers or fragments	Irish continental shelf	PP, PET, PA, AC	Sewage treatment plants	[51]
Fibers	Deep sea and southern ocean	PVC, PES, PA	Wind, seabird	[45]
Resin pellets	North Atlantic ocean and Caribbean sea	HDPE, LDPE, PP, PE	Wind, plankton	[52]
Fragments, sheets, pellets, foam	Cape cod, Massachusetts to the Caribbean sea	PP, HDPP, LDPP PVC, PS, PET	Wind, dust, wave	[53]
Fragments, fishing net, pellet, fibers	Northeast Pacific ocean	HDPP, LDPP, PP PVC, PS, PET	Wind, wave, wear and tear	[54]
Fishing gear, vinyl, rubber	East China sea and South sea of Korea	PA, PP, PES, PE, PVC	Wear and tear, fishing operation	[55]
microfibers	mid-Atlantic, south-western Indian Ocean	PP, PES, AC, viscose	Deep sea organisms	[56]

Table 1.
Classification of microplastic, source of transport, and type of marine environment found.

including mid-ocean islands [42], poles[47], and ocean depths [31]. Pellets are lost during loading and transit, both on land and at sea, as well as during processing at plastic molding plants. Because of their lightness and durability, lost pellets can travel long miles in the ocean before being stranded, either temporarily or permanently [42]. While plastic litter can be found across the whole of the marine ecosystem, the distribution of this debris and microplastics varies in type and nature [16, 46]. Beached litter descriptions frequently mention a variety of plastic manufacture pellets (**Table 1**) [57].

5. Risks of microplastic pollution in the aquatic ecosystem

Plastics have really been recognized as a substantial component of marine plastic pollution for centuries, but their biological and environmental implications on marine ecosystems have only recently been emphasized and appreciated [16, 21]. Microplastics pose a great risk to aquatic life, as their small size makes them readily available to a wide range of marine organisms, and it is of increasing scientific concern (**Figure 4**) [13, 19–21, 34, 45].

Chemical toxins, indigestibility, choking dangers, and a channel for the spread of microbes are just a few of the potential risks that microplastics provide to organisms. These threats increase the risk to aquatic fish and human survival. In this section, we discussed the various risks microplastic pollution posed to the marine ecosystem.

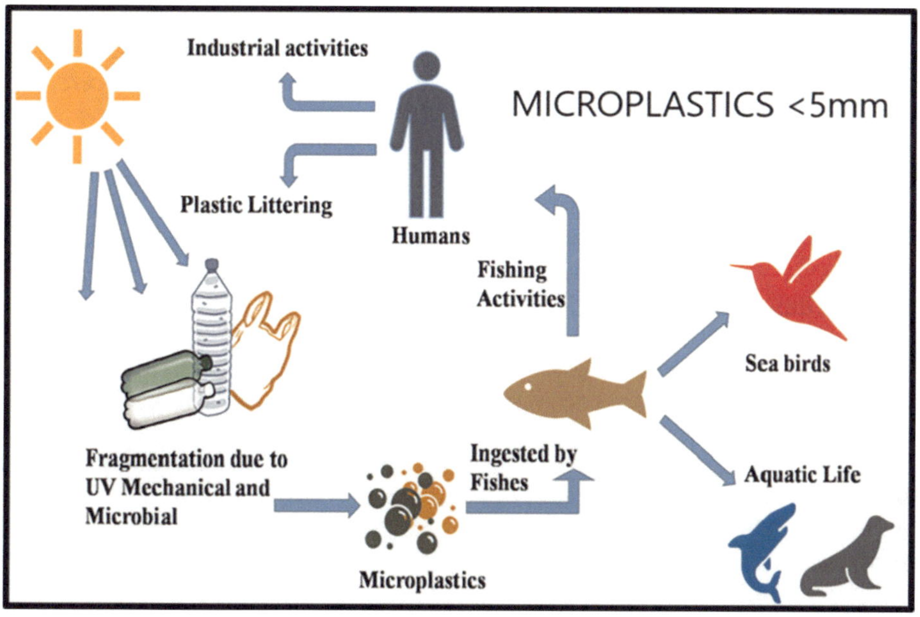

Figure 4.
The effect of microplastic pollution on marine ecosystem [58].

5.1 Chemical toxins

Plastic's durability and corrosion resistance make it an appealing and preferable material to employ, but it also makes it very resistant to deterioration, making the dumping of plastic litter troublesome [45]. The composition of this plastic waste, as well as the enormous surface area of microplastics, makes them vulnerable to attaching watery organic contaminants and hazardous plasticizer leaching. Ingestion of microplastics may thus introduce toxins to the bottom of the food chain in the marine ecosystem, where toxic chemical buildup in the tissues of aquatic living species is possible [59].

Perhaps because plastics are commonly considered to be biochemically inert [59, 60], plastic additives, also known as "plasticizers," may be fully integrated into plastics during manufacturing and injection molding to improve their properties or extend their life by providing resistance to heat (e.g., polybrominated diphenyl ethers), oxidative damage (e.g., nonylphenol), and antimicrobials (triclosan) [37, 61]. These additives are harmful to the environment and the marine ecosystem. Because they both prolong the decomposition time-frames of plastic and may seep out potentially hazardous chemicals into marine aquatic life [45, 62, 63].

A few of these chemicals can move away from the synthetic matrix of plastic due to inadequate polymerization of polymers during manufacture. The extent to which these additives leak from polymers is determined by the pore size of the polymer matrix, which varies by polymer, the additive's size and characteristics, and environmental circumstances, such as weathering [16, 19, 59]. Because microplastics have a high surface-area-to-volume ratio, live species in the marine ecosystem may be directly exposed to leached additives after ingesting microplastics. These chemicals and monomers have the potential to disrupt biologically vital processes, perhaps leading to endocrine disruption, which can have an impact on movement, reproduction and development, and carcinogenesis [45, 62, 64].

Polybrominated diphenyl ethers, phthalates, and the component monomer bisphenol A are well-known endocrine disruptors because they can mimic, compete with, or alter the synthesis of natural hormones [63]. Chemical imbalance can result in temporary or permanent morphological changes in aquatic creatures during their formative phases, as well as sexual disruption in adults. In aquatic invertebrates and fish, phthalates have been related to a variety of molecular and whole-organism consequences, including genotoxic damage (micronuclei and death in mussel hemocytes), restricted motility in invertebrates, and intersex abnormalities in fish [65].

5.2 Indigestibility

While larger forms of garbage are easier to remove from a beach, microplastics are more challenging to eliminate but appear less apparent. Microplastics, due to their microscopic size, have the potential to be consumed by a variety of marine biota [15, 66]. Microplastic consumption in the wild is difficult to observe methodologically [67], however, an increasing number of studies are reporting microplastic ingestion across the food chain. The marine ecological danger associated with microplastics is the increased likelihood of ingestion by animals, such as birds, fish, and invertebrates, resulting in diminished foraging capacity and feeding stimulation, nutritional loss, and gastrointestinal issues [22, 68].

Microplastics pose a significant risk to aquatic life due to their small size, which makes them easily accessible to a wide range of marine creatures, and it is a growing scientific concern [13, 18–21, 31]. In addition to the possible negative effects of swallowing microplastics, toxic responses could emerge from endogenous pollutants leaking from the microplastics and external pollutants adhering to and trying to disassociate from the microplastics. Moreover, utilizing fluorescent nanospheres, phagocytic uptake of nanoplastics in a heterotrophic ciliate was observed. These lower-trophic level creatures are especially susceptible to swallowing microplastics since many of them are indiscriminate feeders with poor ability to distinguish between plastic particles and food particles [16]. As a result, microplastics will be widely and easily available to a wide range of planktonic creatures, including the larval stages of a number of industrially useful species found in the euphotic zone [13, 38]. This interaction between plankton and microplastics is theoretically amplified in gyres, where plankton numbers are low and microplastic intakes are high due to plastic deposition by ocean currents [16].

Microplastics can be consumed by a variety of marine living animals, including seabirds, crustaceans, and fish [69, 70]. Microplastics were found in the intestines of 35% of the planktivorous mesopelagic fish dissected in the north Pacific central gyre [71]. Plastic fibers, pieces, and coatings were also discovered in 13 of 141 mesopelagic fish captured in the north Pacific gyre [72]. In total, 83% of Nephrops sp. sampled in the Clyde sea (Scotland) had consumed pollutants. This economically useful, omnivorous, benthic crustacean primarily ate portions of monofilament line, and plastic bag shards [73].

Plastic fibers in the ecosystem can be as small as one nanometer in diameter and 15 nanometers in length, making them easily accessible to minute planktonic species [74]. Such fibers may be particularly hazardous as they may clump and knot, potentially preventing egestion [73]. In all of the preceding situations, the marine species may have consumed the microplastics intentionally, mistaking them for prey or food. It is yet to be determined whether the consumption of non-polluted microplastics has any substantial detrimental health impacts on biota, such as sickness, death, or reproductive success [75]. Once eaten, microplastics may pose a mechanical hazard to tiny animals, comparable to the consequences reported with macroplastics and bigger species [13, 45].

5.3 Channels for the spread of microbes and adhered pollutants

Marine plastic pollutants, particularly microplastics, are vulnerable to contamination from a variety of waterborne contaminants, including aqueous metals [15, 27], produce harmful chemicals [19], and persistent organic pollutants (POPs), also known as hydrophobic organic contaminants (HOCs) [57]. Such compounds are typically found in the highest quantities in the sea-surface microlayer, which also contains the largest concentrations of low-density microplastics [19, 57, 59].

Under optimal circumstances, phenanthrene was more likely to stick to plastics than to sediments. However, if heavily polluted microplastics come into interaction with non-contaminated sediments, the deposition differential would allow phenanthrene desorption to organic materials in the sediment [5]. A variety of contaminants, including PCBs, PAHs, DDTs and their metabolites, PBDEs, and bisphenol A, were found adhering to the surface of plastic pieces (less than 10 mm) tested from pelagic and neritic stations [76].

Microplastic waste containing POPs may be carried across seas, damaging marine ecosystems [77], or swallowed by marine creatures, transmitting poisons from the environment to aquatic life (i.e., a "Trojan horse" effect) [20, 38]. Many POPs are hazardous, causing endothelial dysfunction, mutagenesis, and/or cancer, and have the potential to biomagnify in higher-trophic organisms [5]. Ref. [78] came to a similar conclusion when they discovered that ingestion of plastic particles hampered the accumulation of fat deposits in migratory red phalaropes (Phalaropus fulicarius), affecting long-distance migration and possibly their reproductive effort on breeding places.

5.4 Choking effect

Plastic pieces and microplastics may also obstruct feeding tentacles and/or impede food transit through the digestive tract [70] or produce pseudo-satiation [the feeling of being full], resulting in reduced food consumption [21, 32]. However, [26, 32] argue that numerous marine organisms have the ability to eliminate foreign particles, such as sediment, natural decaying organic matter, and particulates from their bodies without harm, as illustrated by polychaete worms, which ingested microplastics from their surrounding sediment and then egested them in their fecal contamination casts [20].

Ingestion of plastics may cause blocking of stomach enzyme secretion, decreased eating stimulation, decreased steroid hormone balance, prolonged ovulation, and fertility problems in several marine animals [79]. Ingestion of plastic waste by small fish and seabirds, for example, can limit food intake, induce internal damage, and death due to intestinal infection [22, 23, 80]. Nonetheless, the magnitude of the injury will differ between species. Because of their inability to excrete ingested plastic material, Procellariiformes, for example, are more vulnerable [79, 81].

6. The role of microbes in removal of micro-plastic from marine ecosystem

The environmental problems caused by microplastics in the marine ecosystem are continuously growing [82]. The most common microplastics, also known as synthetic polymers, that are found in the marine ecosystem include PE, PA, PP, PS, PES, AC, PU, HDPP, LDPP, PI, PMMA, PFE, PVC and PVDC [4]. Nevertheless, many conventional plastics, such as PE, PP, PS, PVC, and PET, are not biodegradable, and their increasing accumulation in the ecosystem has posed a danger to the environment [83]. To contend with this man-made challenge, contemporary wastewater treatment facilities need to necessitate fresh technologies [84]. Modern technology provides methods for limiting the availability of microplastics in an aquatic environment. However, such technologies seem to be either inadequate or prohibitively expensive, in addition to being time-consuming in both circumstances. Despite the fact that several microplastic products are considered structural pollutants that do not readily biodegradable or deteriorate at an extremely slow rate, microbial degradation is still a prevalent remediation technique because it is inexpensive and environmentally friendly nature [84, 85].

Microbial degradation can be achieved by using single or connected bio-cultures, including bacteria, algae, and fungi, which have been demonstrated to consume these polymeric materials and generate them into environmentally sustainable carbon compounds. In essence, no microbial techniques can eliminate microplastics from the ecosystem entirely and in an acceptable amount of time [84]. According to research, saturated synthetic polymer chains do not favor microbe degradation, whereas biodegradable polymers incorporate heteroatoms inside the hydrocarbon chains and hence degrade quickly when exposed to favorable weather conditions [86].

The removal rate of microplastic is determined by its creation and the circumstances under which it is exposed, which can range from abiotic factors (wind, waves, heat, and humidity) to microorganism assimilation, such as bacteria, algae, and fungi [87]. As a result, polymer degradation can be categorized as either abiotic or biotic [88]. Abiotic degradation refers to decomposition characterized by factors in the environment, such as temperature, UV irradiation, wind, and waves. Biotic degradation, on the other hand, is defined as the degradation process triggered by the actions of microorganisms that transform and ingest the polymer, modifying its qualities [89].

6.1 Mechanism of biodegradation of microplastics by microbes

The adherence of the microorganism to the surface of the polymer, preceded by the colonization of the external surface, growth of the microbial, use of the polymer as a source of carbon and energy, and final degradation of the polymer is the primary mechanism for microbial degradation [88, 90]. Microorganisms can stick to the surface of a polymer if it is hydrophilic. Once anchored to the surface, the organism can grow by utilizing the polymer as a source of carbon and energy. Polymer biodegradation happens by hydrolysis after colonization; first, the enzyme catalyzes the substrate material and then facilitates the hydrolysis reaction. Polymers degrade into small molecular weight oligomers, dimers, and monomers before finally mineralization to CO_2 and H_2O [83]. The surface composition can quantify the scope of colonization on the polymer, as hydrophilic areas are much more conveniently colonized by microbes. This is a restriction since the polymer's water-repellent surface contradicts the porous structure of the microorganisms (**Figure 5**) [89].

6.2 Biotic degradation

Microplastic biodegrades as a consequence of degradation by microbes in the marine environment. However, because of their size, macroplastics (larger plastic debris) do not make the optimum source of nutrients for biotic degrading agents;

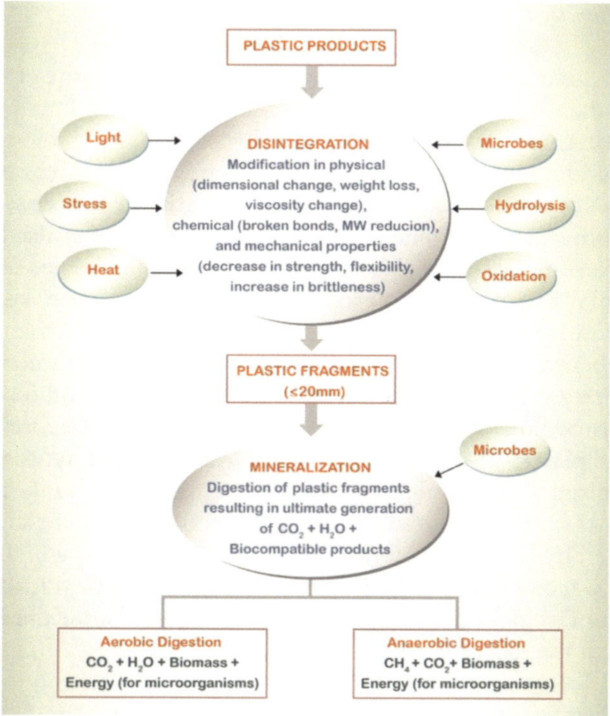

Figure 5.
Mechanism of microbial degradation of microplastic [91].

either the enzymes secreted by the microbes are insufficient to denature the macroplastics, or they contain not easily and quickly biodegradable for biological cell uptake [29]. Synthetic polymer plastics must first be changed into carbon molecules prior to being mineralized by microbial pathogens during the degradation reaction. Plastics' (polymers') organic molecules' size is bigger than the particle sizes of a microorganism's cellular membrane. As a result, they must be metabolized into tiny pieces before being assimilated and biodegraded within microbial cells. As a result, finer particles of plastic created as the result of environmental factors degradation are of sufficient size to be broken down even more by microbial cells [92]. Bacteria, fungi, and algae are the most common microorganisms found in marine ecosystems.

Microbial enzymes are responsible for biotic degradation. Chemical compounds are converted into simplified chemical compounds, metabolized, and deposited in primary-level cycles, such as carbon, nitrogen, and sulfur through microbial degradation. Carbon dioxide, methane, and microbial extracellular matrix components are among the by-products of this system [93, 94]. Microbial character traits, such as microbe form, propagation, developmental stage (temperature, pH, availability of oxygen, essential minerals, etc.), and enzymatic categories (intracellular and/or extracellular enzymes contributing to exo or endo polymer cleaving). Surface conditions (size, water-soluble, and hydrophilicity properties), first-order frameworks (chemical composition, molecular mass, and molecular dissemination), and relatively high structures (thermodynamic stability, melting temperature, fracture toughness, crystalline structure, and degree of crystallinity) are among the chemical and physical properties of polymers [83].

6.2.1 Biodegradation by bacteria

Many bacteria genera that are commonly found in the marine environment like Bacillus species (e.g., Bacillus subtilis and Bacillus cereus, and Bacillus megaterium), Brevibacillus, Streptomyces, Amycolatopsis, Clostridium, Methanosarcina barkei, Schlegelella, Pseudomonas aeruginosa, Azotobacter spp., Alcanivorax, Hyphomonas, and Cycloclasticus species, Rhodococcus ruber, Serratia marcescens, Staphylococcus aureus, and Streptococcus pyogenes, and other bacterial strains also lead to the microbial degradation of plastics [29, 95–102]. The Bacillus species were discovered to secrete extracellular hydrolytic enzymes, such as lipase, xylanase, keratinase, chitinase, and protease, which resulted in the biodegradation of microplastics [103]. Methanosarcina barkei bacteria strain can degrade the most commonly used plastic polymer, PVC. They can stick to the surface of PVC surfaces and discharge exopolymeric compounds to produce a biofilm, preceded by the discharge of enzymes to breakdown the plastic through enzymatic hydrolysis of the synthetic polymer bonds which resulted in the biodegradation of PVC [104, 105]. Likewise, Rhodococcus ruber will also degrade PE by producing an enzyme laccase, which ultimately resulted in PE degradation [106]. Azotobacter spp., which releases hydroquinone peroxidase, could also degrade PS. PET can also be degraded by Alcanivorax, Hyphomonas, and Cycloclasticus species, which could also alter the physiochemical properties through the use of ester bond hydrolysis [107].

6.2.2 Biodegradation by fungi

Many fungal genera, such as Acremonium, Zalerion maritimum, o Curvularia sp., Cladosporium, Debaryomyces, Emericellopsis, Eupenicillium, Fusarium, Mucor, Paecilomyces, Pullularia, Rhodosporidium, Verticillium, Aspergillus sp., Aureobasidium, Chaetomium, Cryptococcus, Fusarium, Rhizopus arrhizus, Trichoderma, Penicillium sp., Thermoascus, Tritirachium album, Humicola insolens, Rhodotorula aurantiaca, and Kluyveromyces sp. [83, 108–112] also contribute to the microbial degradation of plastics. It has been demonstrated that Aspergillus clavatus can biodegrade LDPE [113]. Zalerion maritimum, the ocean's dominant fungal species, could also degrade PE [114]. The main mechanism of plastic degradation by fungi, such as bacteria, involves fungi adhering to the polymer surface, in which they grow to create a biofilm and produce enzymes that degrade the carbon-carbon bonds occurring in the plastic. The above enzymes have the potential to accelerate the oxidation process as well as degrade plastic into tiny pieces (e.g., oligomers, dimers, and monomers). For example, fungi found in marine habitats, such as Penicillium citrinum and Fusarium oxysporum, breakdown PET, and Trichoderma harzianum release manganese peroxidase, lignin peroxidase, and laccase that breakdown PE and PU [114].

6.2.3 Biodegradation by algae

Algae are frequently used throughout tested microorganisms for investigating the harmful effects of microplastics. However, various algae, both photorespiration and heterotrophic, have been extensively researched for their key responsibilities in the microbial degradation of microplastics [84, 85]. They are capable of removing both inorganic and organic contaminants from a diverse range of environments by soaking up, removing impurities, or metabolizing them into healthy and safe levels [115, 116].

They colonize the outer layer of microplastics by secreting extracellular polymeric compounds, and this colonization could well result in effectual deterioration. The existence of polymeric materials, as well as plastic wastes, encourages the generation of extracellular polymeric compounds [117]. Several algal species are effective at microbial degradation of microplastics. These include Phormidium lucidum, Oscillatoria subbrevis, Scenedesmus dimorphus, diatom Navicula pupula, Chlorella, Spirogyra, Nostoc, Spirulina sp., Anabaena spiroides, and Navicula pupula [118–120]. Bioactive compounds produced by some algae have been found to biodegrade microplastics. Phormidium lucidum and Oscillatoria subbrevis, for example, can break down easily PE and LDPE [121]. Discostella spp., Navicula spp., Amphora spp., and Fragilaria spp. algal biofilms have been discovered to deplete LDPE, PP, and PET in the marine ecosystem [122]. After forming a biofilm on the plastic surface, algae use the carbon available on the plastic as a feed ingredient, softening and lessening the plastic. Furthermore, species can produce extracellular polymeric compounds and enzymes, such as PETase, which degrade PET [123]. Plastic degradation by algae remains in its early stages and requires more research.

7. Conclusion and recommendation

7.1 Conclusion

Plastic pollution in the marine ecosystem is a growing concern due to the negative effects it has on aquatic habitats. Microplastic pollution has become a serious global issue that has a detrimental effect on the food chain in the marine ecosystem. The main sources of microplastic pollution in the marine ecosystem have been identified to result from general littering, plastic waste mismanagement, fishing gears, synthetic textiles, marine coatings, personal care products, plastic pellets, city dust, and release of wastewater from sewage treatment plants. This is the outcome of indiscriminate waste dumping, which is either directly or indirectly transmitted to our seas and oceans. Because microplastics are the same size as prey and are mistaken for food, they pose a threat to many marine organisms. When swallowed, it has a negative impact on marine organisms, facilitating the transmission of artificial chemicals or hydrophobic watery toxins to aquatic life. Microplastic pollution has contaminated various drinking sources, salt water, and other regularly consumed foods. Chemical toxication, indigestibility, choking of marine ecosystems, and a pathway for microbial propagation are all negative effects of microplastic contamination on the marine environment. Furthermore, the effects of microplastic pollution vary from the molecular level of an organism to its physiological mechanisms and include bad organism health and poor economic services. These threats increase the risk to aquatic fish's and human survival. Significant awareness about the harmful effects of microplastics has prompted some regions of the world, including the United Kingdom, the United States, and Canada, to take action. These initiatives have focused almost entirely on prohibiting the use of microbeads in various items, such as personal care and skincare products.

7.2 Recommendation

Microplastics have been found to be consumed by a variety of marine organisms in laboratory and field research. More research is needed to determine whether microplastic consumption alone causes unfavorable health impacts, such as mortality, morbidity, and reproductive success, or whether such a contaminant can be consistently transferred up the food chain in the marine ecosystem. Toxic chemical transfer to biota via microplastic intake is a major concern. However, just a few studies have reported on toxicity investigations, including microplastic vectors. More quantitative research should be conducted to investigate the toxins [toxic chemicals] transfer of microplastics to marine species, as well as any possible dangers of transfer from consumable marine organisms to people.

The most pressing need in this subject is to raise public understanding about the inert impacts of microplastics. This would encourage numerous inventions aimed at reducing the use and consumption of plastic and its byproducts. The most essential way to reduce plastic entry into the ecosystem is to gather and reuse plastic particles. To avert future threats, the best answer is to discontinue production and seek alternatives to plastic items.

Author details

Paul Agbekpornu[1*] and Isaac Kevudo[2]

1 Department of Statistics and Actuarial Science, Kwame Nkrumah, University of Science and Technology (KNUST), Ghana

2 School of Medicine, University of Health and Allied Science (UHAS), Ghana

*Address all correspondence to: pagbekpornu@aims.edu.gh

References

[1] Desidery L, Lanotte M. Polymers and plastics: Types, properties, and manufacturing. In: Plastic Waste for Sustainable Asphalt Roads. Woodhead Publishing; 2022. pp. 3-28

[2] Aamer AS, Fariha H, Abdul H, Safia A. Biological degradation of plastics: A comprehensive review. Biotechnology Advances. 2008;**26**(3):246-265

[3] Andrady AL, Neal MA. Applications and societal benefits of plastics. Philosophical Transactions of the Royal Society, B: Biological Sciences. 2009;**364**: 1977-1984

[4] Geyer R, Jambeck JR, Law KL. Production, use, and fate of all plastics ever made. Science Advances. 2017;**3**(7): e1700782

[5] Cole M, Lindeque P, Halsband C, Galloway TS. Microplastics as contaminants in the marine environment: A review. Marine Pollution Bulletin. 2011;**62**:2588-2597

[6] Rillig MC. Microplastic in Terrestrial Ecosystems and the Soil. ACS Publications. 2012. pp. 6453-6454

[7] Facts PEPT. An Analysis of European Plastics Production, Demand and Waste Data. Plastics Europe. 2019

[8] Wagner M, Scherer C, Alvarez-Muňoz D, Brennholt N, Bourrain X, Buchinger S, et al. Microplastics in freshwater ecosystems: What we know and what we need to know. Environmental Sciences Europe. 2014;**26**(1):1-9

[9] Jambeck JR, Geyer R, Wilcox C, Siegler TR, Perryman M, Andrady A, et al. Plastic waste inputs from land into the ocean. Science. 2015;**347**:768-771

[10] Meijer LJ, van Emmerik T, van der Ent R, Schmidt C, Lebreton L. More than 1000 rivers account for 80% of global riverine plastic emissions into the ocean. Science Advances, 2021;**7**(18):eaaz5803. OurWorldInData.org/plastic-pollution

[11] Duis K, Coors A. Microplastics in the aquatic and terrestrial environment: Sources (with a specific focus on personal care products), fate and effects. Environmental Sciences Europe. 2016;**28** (1):1-25

[12] Chatterjee S, Sharma S. Microplastics in our oceans and marine health. Field Actions Science Reports. The Journal of Field Actions. 2019;**19**:54-61

[13] Fendall LS, Sewell MA. Contributing to marine pollution by washing your face: Microplastics in facial cleansers. Marine Pollution Bulletin. 2009;**58**:1225-1228

[14] Gregory MR. Environmental implications of plastic debris in marine settings: Entanglement, ingestion, smothering, hangers-on, hitch-hiking and alien invasions. Philosophical Transactions of the Royal Society, B: Biological Sciences. 2009;**364**:2013-2025

[15] Betts K. Why small plastic particles may pose a big problem in the oceans. Environmental Science & Technology. 2008;**42**:8995

[16] Moore CJ. Synthetic polymers in the marine environment: A rapidly increasing, long-term threat. Environmental Research. 2008;**108**: 131-139

[17] Arthur C, Bamford H, Baker J. The occurrence, effects and fate of small plastic debris in the oceans. In: Proceedings of the International

Research Workshop on the Occurrence. Tacoma, WA, USA: Effects and Fate of Microplastic Marine Debris; 2008. pp. 9-11

[18] Barnes DKA. Biodiversity: Invasions by marine life on plastic debris. Nature. 2002;**416**:808-809

[19] Ng KL, Obbard JP. Prevalence of microplastics in Singapore's coastal marine environment. Marine Pollution Bulletin. 2006;**52**:761-767

[20] Thompson RC, Olsen Y, Mitchell RP, Davis A, Rowland SJ, John AWG, et al. Lost at sea: Where is all the plastic? Science. 2004;**2004**:838

[21] Derraik JGB. The pollution of the marine environment by plastic debris: A review. Marine Pollution Bulletin. 2002; **44**:842-852

[22] Ryan PG, Moore CJ, van Franeker JA, Moloney CL. Monitoring the abundance of plastic debris in the marine environment. Philosophical Transactions of the Royal Society, B: Biological Sciences. 2009;**364**:1999-2012

[23] Carpenter EJ, Smith KL Jr. Plastics on the Sargasso Sea surface. Science. 1972;**175**:1240-1241

[24] Sutherland WJ, Clout M, Côté IM, Daszak P, Depledge MH, Fellman L, et al. A horizon scan of global conservation issues for 2010. Trends in Ecology & Evolution. 2010;**25**:1-7

[25] Rands MRW, Adams WM, Bennun L, Butchart SHM, Clements A, Coomes D, et al. Biodiversity conservation: Challenges beyond 2010. Science. 2010; **329**:1298-1303

[26] Andrady AL. Microplastics in the marine environment. Marine Pollution Bulletin. 2011;**62**:1596-1605

[27] Ashton K, Holmes L, Turner A. Association of metals with plastic production pellets in the marine environment. Marine Pollution Bulletin. 2010;**60**:2050-2055

[28] Boucher J, Friot D. Primary Microplastics in the Oceans: A Global Evaluation of Sources. Vol. 10. Gland, Switzerland: Iucn; 2017

[29] Haque F, Fan C. Microplastics in the Marine Environment: A Review of Their Sources, Formation, Fate, and Ecotoxicological Impact. IntechOpen; 2022

[30] Long Z, Pan Z, Wang W, Ren J, Yu X, Lin L, et al. Microplastic abundance, characteristics, and removal in wastewater treatment plants in a coastal city of China. Water Research. 2019;**155**: 255-265

[31] Lozano RL, Mouat J. Marine Litter in the North-East Atlantic Region: Assessment and Priorities for Response. London, United Kingdom: OSPAR Commission; 2009. p. 127

[32] Thompson RC. Plastic debris in the marine environment: Consequences and solutions. In: Krause JC, Nordheim H, Bräger S, editors. Marine Nature Conservation in Europe. Stralsund, Germany: Federal Agency for Nature Conservation; 2006. pp. 107-115

[33] Leslie HA. Plastic in Cosmetics: Are we polluting the environment through our personal care? UNEP. 2015;**2015**:33

[34] Lassen C, Foss Hansen S, Magnusson K, Noren F, Bloch Hartmann NI, Rehne Jensen P, et al. Microplastics: Occurrence, Effects and Sources of Releases to the Environment in Denmark (The Danish Environmental Protection Agency). 2015

[35] Zhongming Z, Linong L, Xiaona Y, Wangqiang Z, Wei L. Microplastic Sizes in Hudson-Raritan Estuary and Coastal Ocean Revealed. 2021

[36] Driedger AGJ, Dürr HH, Mitchell K, Van Cappellen P. Plastic debris in the Laurentian Great Lakes: A review. Journal of Gt Lakes Research. 2015;**41**:9-19. DOI: 101016/jjglr2014.12.020

[37] Browne MA, Galloway T, Thompson R. Microplastic – an emerging contaminant of potential concern? Integrated Environmental Assessment and Management. 2007;**3**:559-561

[38] Gregory MR. Plastic 'scrubbers' in hand cleansers: A further (and minor) source for marine pollution identified. Marine Pollution Bulletin. 1996;**32**:867-871

[39] Brillant M, MacDonald B. Postingestive selection in the sea scallop (Placopecten magellanicus) on the basis of chemical properties of particles. Marine Biology. 2002;**141**:457-465

[40] Moore CJ, Moore SL, Weisberg SB, Lattin GL, Zellers AF. A comparison of neustonic plastic and zooplankton abundance in southern California's coastal waters. Marine Pollution Bulletin. 2002;**44**:1035-1038

[41] Pruter AT. Sources, quantities and distribution of persistent plastics in the marine environment. Marine Pollution Bulletin. 1987;**18**:305-310

[42] Ivar do Sul JA, Spengler A, Costa MF. Here, there and everywhere, small plastic fragments and pellets on beaches of Fernando de Noronha (EquatorialWestern Atlantic). Marine Pollution Bulletin. 2009;**58**:1236-1238

[43] Mato Y, Isobe T, Takada H, Kanehiro H, Ohtake C, Kaminuma T. Plastic resin pellets as a transport medium for toxic chemicals in the marine environment. Environmental Science & Technology. 2001;**35**:318-324

[44] Essel R, Engel L, Carus M, Ahrens RH. Sources of microplastics relevant to marine protection in Germany. Text. 2015;**64**(2015):1219-1226

[45] Barnes DKA, Galgani F, Thompson RC, Barlaz M. Accumulation and fragmentation of plastic debris in global environments. Philosophical Transactions of the Royal Society, B: Biological Sciences. 2009;**364**:1985-1998

[46] Martinez E, Maamaatuaiahutapu K, Taillandier V. Floating marine debris surface drift: Convergence and accumulation toward the South Pacific subtropical gyre. Marine Pollution Bulletin. 2009;**58**:1347-1355

[47] Barnes DKA, Walters A, Gonçalves L. Macroplastics at sea around Antarctica. Marine Environmental Research. 2010;**70**:250-252

[48] Woodall LC, Sanchez-Vidal A, Canals M, Paterson GL, Coppock R, Sleight V, et al. The deep sea is a major sink for microplastic debris. Royal Society Open Science. 2014;**1**(4):140317

[49] Browne MA, Galloway TS, Thompson RC. Spatial patterns of plastic debris along estuarine shorelines. Environmental Science & Technology. 2010;**44**(9):3404-3409

[50] Browne MA, Crump P, Niven SJ, Teuten E, Tonkin A, Galloway T, et al. Accumulation of microplastic on shorelines woldwide: Sources and sinks. Environmental Science & Technology. 2011;**45**(21):9175-9179

[51] Martin J, Lusher A, Thompson RC, Morley A. The deposition and

accumulation of microplastics in marine sediments and bottom water from the Irish continental shelf. Scientific Reports. 2017;7(1):1-9

[52] Law KL, Morét-Ferguson S, Maximenko NA, Proskurowski G, Peacock EE, Hafner J, et al. Plastic accumulation in the North Atlantic subtropical gyre. Science. 2010;**329** (5996):1185-1188

[53] Morét-Ferguson S, Law KL, Proskurowski G, Murphy EK, Peacock EE, Reddy CM. The size, mass, and composition of plastic debris in the western North Atlantic Ocean. Marine Pollution Bulletin. 2010;**60**(10):1873-1878

[54] Doyle MJ, Watson W, Bowlin NM, Sheavly SB. Plastic particles in coastal pelagic ecosystems of the Northeast Pacific ocean. Marine Environmental Research. 2011;**71**(1):41-52

[55] Lee DI, Cho HS, Jeong SB. Distribution characteristics of marine litter on the sea bed of the East China Sea and the South Sea of Korea. Estuarine, Coastal and Shelf Science. 2006;**70**(1–2): 187-194

[56] Taylor ML, Gwinnett C, Robinson LF, Woodall LC. Plastic microfibre ingestion by deep-sea organisms. Scientific Reports. 2016;**6**(1):1-9

[57] Rios LM, Moore C, Jones PR. Persistent organic pollutants carried by synthetic polymers in the ocean environment. Marine Pollution Bulletin. 2007;**54**:1230-1237

[58] Issac MN, Kandasubramanian B. Effect of microplastics in water and aquatic systems. Environmental Science and Pollution Research. 2021;**28**(16): 19544-19562

[59] Teuten EL, Saquing JM, Knappe DRU, Barlaz MA, Jonsson S, BjÃrn A, et al. Transport and release of chemicals from plastics to the environment and to wildlife. Philosophical Transactions of the Royal Society, B: Biological Sciences. 2009;**364**:2027-2045

[60] Roy PK, Hakkarainen M, Varma IK, Albertsson A-C. Degradable polyethylene: Fantasy or reality. Environmental Science & Technology. 2011;**45**:4217-4227

[61] Thompson RC, Swan SH, Moore CJ, vom Saal FS. Our plastic age. Philosophical Transactions of the Royal Society, B: Biological Sciences. 2009b; **364**:1973-1976

[62] Lithner D, Larsson Å, Dave G. Environmental and health hazard ranking and assessment of plastic polymers based on chemical composition. Science of the Total Environment. 2011;**409**:3309-3324

[63] Talsness CE, Andrade AJM, Kuriyama SN, Taylor JA, vom Saal FS. Components of plastic: Experimental studies in animals and relevance for human health. Philosophical Transactions of the Royal Society, B: Biological Sciences. 2009;**364**:2079-2096

[64] Lithner D, Damberg J, Dave G, Larsson A. Leachates from plastic consumer products - Screening for toxicity with Daphnia magna. Chemosphere. 2009;**74**:1195-1200

[65] Oehlmann J, Schulte-Oehlmann U, Kloas W, Jagnytsch O, Lutz I, Kusk KO, et al. A critical analysis of the biological impacts of plasticizers on wildlife. Philosophical Transactions of the Royal Society, B: Biological Sciences. 2009;**364**: 2047-2062

[66] Thompson RC, Moore CJ, vom Saal FS, Swan SH. Plastics, the environment and human health: Current consensus

and future trends. Philosophical Transactions of the Royal Society, B: Biological Sciences. 2009a;**364**:2153-2166

[67] Browne MA, Dissanayake A, Galloway TS, Lowe DM, Thompson RC. Ingested microscopic plastic translocates to the circulatory system of the mussel, Mytilus edulis (L.). Environmental Science & Technology. 2008;**42**:5026-5031

[68] Graham ER, Thompson JT. Deposit- and suspension-feeding sea cucumbers (Echinodermata) ingest plastic fragments. Journal of Experimental Marine Biology and Ecology. 2009;**368**:22-29

[69] Blight LK, Burger AE. Occurrence of plastic particles in seabirds from the eastern North Pacific. Marine Pollution Bulletin. 1997;**34**:323-325

[70] Tourinho PS, Ivar Do Sul JA, Fillmann G. Is marine debris ingestion still a problem for the coastal marine biota of southern Brazil? Marine Pollution Bulletin. 2010;**60**:396-401

[71] Boerger CM, Lattin GL, Moore SL, Moore CJ. Plastic ingestion by planktivorous fishes in the North Pacific Central Gyre. Marine Pollution Bulletin. 2010;**60**:2275-2278

[72] Davison P, Asch RG. Plastic ingestion by mesopelagic fishes in the North Pacific Subtropical Gyre. Marine Ecology Progress Series. 2011;**432**:173-180

[73] Murray F, Cowie PR. Plastic contamination in the decapod crustacean Nephrops norvegicus (Linnaeus, 1758). Marine Pollution Bulletin. 2011;**62**:1207-1217

[74] Frias JPGL, Sobral P, Ferreira AM. Organic pollutants in microplastics from two beaches of the Portuguese coast. Marine Pollution Bulletin. 2010;**60**:1988-1992

[75] Zarfl C, Fleet D, Fries E, Galgani F, Gerdts G, Hanke G, et al. Microplastics in oceans. Marine Pollution Bulletin. 2011;**62**:1589-1591

[76] Hirai H, Takada H, Ogata Y, Yamashita R, Mizukawa K, Saha M, et al. Organic micropollutants in marine plastics debris from the open ocean and remote and urban beaches. Marine Pollution Bulletin. 2011;**62**:1683-1692

[77] Zarfl C, Matthies M. Are marine plastic particles transport vectors for organic pollutants to the Arctic? Marine Pollution Bulletin. 2010;**60**:1810-1814

[78] Connors PG, Smith KG. Oceanic plastic particle pollution: Suspected effect on fat deposition in red phalaropes. Marine Pollution Bulletin. 1982;**13**(1):18-20

[79] Azzarello MY, Van Vleet ES. Marine birds and plastic pollution. Marine Ecology Progress Series. 1987;**37**(2/3):295-303

[80] Zitko V, Hanlon M. Another source of pollution by plastics: Skin cleansers with plastic scrubbers. Marine Pollution Bulletin. 1991;**22**:41-42

[81] Furness RW. Ingestion of plastic particles by seabirds at Gough Island, South Atlantic Ocean. Environmental Pollution Series A, Ecological and Biological. 1985;**38**(3):261-272

[82] Ghosh SK, Pal S, Ray S. Study of microbes having potentiality for biodegradation of plastics. Environmental Science and Pollution Research. 2013;**20**(7):4339-4355

[83] Tokiwa Y, Calabia BP, Ugwu CU, Aiba S. Biodegradability of plastics.

International Journal of Molecular Sciences. 2009;**10**(9):3722-3742

[84] Miloloža M, Cvetnić M, Kučić Grgić D, Ocelić Bulatović V, Ukić Š, Rogošić M, et al. Biotreatment strategies for the removal of microplastics from freshwater systems: A review. Environmental Chemistry Letters. 2022; **2022**:1-26

[85] Amobonye A, Bhagwat P, Singh S, Pillai S. Plastic biodegradation: Frontline microbes and their enzymes. Science of the Total Environment. 2021;**759**:143536

[86] Raghavan D. Characterization of biodegradable plastics. Polymer-Plastics Technology and Engineering. 1995;**34** (1):41-63

[87] Karamanlioglu M, Preziosi R, Robson GD. Abiotic and biotic environmental degradation of the bioplastic polymer poly (lactic acid): A review. Polymer Degradation and Stability. 2017;**137**:122-130

[88] Oliveira J, Belchior A, da Silva VD, Rotter A, Petrovski Ž, Almeida PL, et al. Marine environmental plastic pollution: Mitigation by microorganism degradation and recycling valorization. Frontiers in Marine Science. 2020;**7**: 567126

[89] Restrepo-Flórez JM, Bassi A, Thompson MR. Microbial degradation and deterioration of polyethylene–A review. International Biodeterioration & Biodegradation. 2014;**88**:83-90

[90] Alshehrei F. Biodegradation of synthetic and natural plastic by microorganisms. Journal of Applied & Environmental Microbiology. 2017;**5**(1): 8-19

[91] Krzan A, Hemjinda S, Miertus S, Corti A, Chiellini E. Standardization and certification in the area of environmentally degradable plastics. Polymer Degradation and Stability. 2006;**91**(12):2819-2833

[92] Zhang K, Hamidian AH, Tubić A, Zhang Y, Fang JK, Wu C, et al. Understanding plastic degradation and microplastic formation in the environment: A review. Environmental Pollution. 2021;**274**:116554

[93] Banker D, Ahuja A, Sanadhya H, Purohit C, Chandra S, Sethi R, et al. Clinical performance of biodegradable polymer-coated sirolimus-eluting stents in unselected real-world population with coronary artery disease: Results from the multicenter CORE Registry. Minerva Cardioangiologica. 2015;**64**(1):9-14

[94] Chandra RUSTGI, Rustgi R. Biodegradable polymers. Progress in Polymer Science. 1998;**23**(7):1273-1335

[95] Nakamiya K, Ooi T, Kinoshita S. Non-heme hydroquinone peroxidase from Azotobacter beijerinckii HM121. Journal of Fermentation and Bioengineering. 1997;**84**(1):14-21

[96] Jeon HJ, Kim MN. Functional analysis of alkane hydroxylase system derived from Pseudomonas aeruginosa E7 for low molecular weight polyethylene biodegradation. International Biodeterioration & Biodegradation. 2015;**103**:141-146

[97] Auta HS, Emenike CU, Jayanthi B, Fauziah SH. Growth kinetics and biodeterioration of polypropylene microplastics by Bacillus sp. and Rhodococcus sp. isolated from mangrove sediment. Marine Pollution Bulletin. 2018;**127**:15-21

[98] Giacomucci L, Raddadi N, Soccio M, Lotti N, Fava F. Polyvinyl chloride biodegradation by Pseudomonas

citronellolis and Bacillus flexus. New Biotechnology. 2019;**52**:35-41

[99] Muhonja CN, Makonde H, Magoma G, Imbuga M. Biodegradability of polyethylene by bacteria and fungi from Dandora dumpsite Nairobi-Kenya. PLoS One. 2018;**13**(7):e0198446

[100] Kyaw BM, Champakalakshmi R, Sakharkar MK, Lim CS, Sakharkar KR. Biodegradation of low density polythene (LDPE) by Pseudomonas species. Indian Journal of Microbiology. 2012;**52**(3): 411-419

[101] Ho BT, Roberts TK, Lucas S. An overview on biodegradation of polystyrene and modified polystyrene: The microbial approach. Critical Reviews in Biotechnology. 2018;**38**(2): 308-320

[102] Arefian M, Tahmourespour A, Zia M. Polycarbonate biodegradation by newly isolated Bacillus strains. Archives of Environmental Protection. 2020;**46** (1):14-20

[103] Chandra P, Singh DP. Microplastic degradation by bacteria in aquatic ecosystem. In: Microorganisms for Sustainable Environment and Health. Elsevier; 2020. pp. 431-467

[104] Nguyen V, Karunakaran E, Collins G, Biggs CA. Physicochemical analysis of initial adhesion and biofilm formation of Methanosarcina barkeri on polymer support material. Colloids and Surfaces B: Biointerfaces. 2016;**143**:518-525

[105] Reisser J, Shaw J, Hallegraeff G, Proietti M, Barnes DK, Thums M, et al. Millimeter-sized marine plastics: A new pelagic habitat for microorganisms and invertebrates. PLoS One. 2014;**9**(6): e100289

[106] Santo M, Weitsman R, Sivan A. The role of the copper-binding enzyme–laccase–in the biodegradation of polyethylene by the actinomycete Rhodococcus ruber. International Biodeterioration & Biodegradation. 2013;**84**:204-210

[107] Denaro R, Aulenta F, Crisafi F, Di Pippo F, Viggi CC, Matturro B, et al. Marine hydrocarbon-degrading bacteria breakdown poly (ethylene terephthalate) (PET). Science of the Total Environment. 2020;**749**:141608

[108] Motta O, Proto A, De Carlo F, De Caro F, Santoro E, Brunetti L, et al. Utilization of chemically oxidized polystyrene as co-substrate by filamentous fungi. International Journal of Hygiene and Environmental Health. 2009;**212**(1):61-66

[109] Das MP, Kumar S. Microbial deterioration of low density polyethylene by Aspergillus and Fusarium sp. International Journal of Chemical Technical Research. 2014;**6**(1): 299-305

[110] Alariqi SA, Kumar AP, Rao BSM, Singh RP. Biodegradation of γ-sterilised biomedical polyolefins under composting and fungal culture environments. Polymer Degradation and Stability. 2006;**91**(5):1105-1116

[111] Ojha N, Pradhan N, Singh S, Barla A, Shrivastava A, Khatua P, et al. Evaluation of HDPE and LDPE degradation by fungus, implemented by statistical optimization. Scientific Reports. 2017;**7**(1):1-13

[112] Pečiulytė D. Microbial colonization and biodeterioration of plasticized polyvinyl chloride plastics. Ekologija. 2002;**4**:7-15

[113] Mor R, Sivan A. Biofilm formation and partial biodegradation of polystyrene by the actinomycete

Rhodococcus ruber. Biodegradation. 2008;**19**(6):851-858

[114] Zeghal E, Vaksmaa A, Vielfaure H, Boekhout T, Niemann H. The potential role of marine fungi in plastic degradation–A review. Frontiers in Marine Science. 2021;**8**:738877

[115] Hwang JH, Sadmani A, Lee SJ, Kim KT, Lee WH. Microalgae: An eco-friendly tool for the treatment of wastewaters for environmental safety. In: Bioremediation of Industrial Waste for Environmental Safety. Singapore: Springer; 2020. pp. 283-304

[116] Hoffmann L, Eggers SL, Allhusen E, Katlein C, Peeken I. Interactions between the ice algae Fragillariopsis cylindrus and microplastics in sea ice. Environment International. 2020;**139**: 105697

[117] Song C, Liu Z, Wang C, Li S, Kitamura Y. Different interaction performance between microplastics and microalgae: The bio-elimination potential of Chlorella sp. L38 and Phaeodactylum tricornutum MASCC-0025. Science of the Total Environment. 2020;**723**:138146

[118] Kumar RV, Kanna GR, Elumalai S. Biodegradation of polyethylene by green photosynthetic microalgae. Journal of Bioremediation & Biodegradation. 2017; **8**(381):2

[119] Hadiyanto H, Khoironi A, Dianratri I, Suherman S, Muhammad F, Vaidyanathan S. Interactions between polyethylene and polypropylene microplastics and Spirulina sp. microalgae in aquatic systems. Heliyon. 2021;**7**(8):e07676

[120] Sarmah P, Rout J. Efficient biodegradation of low-density polyethylene by cyanobacteria isolated from submerged polyethylene surface in domestic sewage water. Environmental Science and Pollution Research. 2018;**25** (33):33508-33520

[121] Chia WY, Tang DYY, Khoo KS, Lup ANK, Chew KW. Nature's fight against plastic pollution: Algae for plastic biodegradation and bioplastics production. Environmental Science and Ecotechnology. 2020;**4**:100065

[122] Smith IL, Stanton T, Law A. Plastic habitats: Algal biofilms on photic and aphotic plastics. Journal of Hazardous Materials Letters. 2021;**2**:100038

[123] Ali SS, Elsamahy T, Al-Tohamy R, Zhu D, Mahmoud YAG, Koutra E, et al. Plastic wastes biodegradation: Mechanisms, challenges and future prospects. Science of the Total Environment. 2021;**780**:146590

Chapter 7

Micro Plastic Challenges in River Delimi Due to Its Interaction with Physicochemical Parameters

Terwase Wuave and Ahmed Sabo

Abstract

Physicochemical interaction with micro plastic at River Delimi represents one of the most pressing threats to water and plant resources as it is a challenge to human health. Micro plastics in the food chain constitutes a major threat to water, soil and plants which affect food safety, it affects public health when consuming products that have this pollutant, such as fruits, and vegetables. Micro plastics (MPs) are tiny particles broken down from larger pieces of plastics, accumulated in River Delimi. This study aimed at assessing the physicochemical parameters (Cd, Zn, Co, Pb and Ni) in water and (Ba, Ce, Rb, La, Nd, Ta, Sm, Sc and Th) while mercury and silver show high concentration (Hg, Ag) plants as it interaction with micro plastics (majorly polyethylene terephthalate (PET) concentration and plasticized polyvinylchloride (PVC both dry and rainy season)in the study area, along Rivers Delimi in Jos North LGA of Plateau State. Water and plant samples were collected from six (6) different irrigation sites in the area, namely Delimi village, Delimi area, Marhaba Masjd, Malam Adamu and Tudun Wada along the River Delimi, Barakin Kogi, and Baraki Naraguta along River Bonga, contain micro particles in edible plants due to the contamination of water-irrigated soils and methods to determine them. The negative effect of micro plastics on various food products and their interaction with physicochemical properties especially heavy metals impact on the environment is determined. Samples were analyzed to determine the concentration of the physicochemical parameters using the Atomic Absorption Spectrometer. Micro plastic are carriers for physicochemical parameters and exhibit diverse interactive effects, these interactions are poorly understood especially how they pose risks on living organisms. The challenges of their combined toxic effects and the potential hazards to human health were also discussed. Results show that the concentration of the physicochemical parameters in water as electrical conductivity, hydrocabonate, chloride, sodium carcium, chromium and lead while plasticized polyvinylchloride (PVC) both dry and rainy season, in river Delimi and river Bonga were pronounce. In plants potassium, phosphorus, molybdenum, manganese, mercury and silver were also high.

Keywords: micro plastics, heavy metals, interactions, microbe, aquatic environment

1. Introduction

1.1 Micro plastics

Additives increase specific properties of plastic polymers; those called butyl tins stabilize polyvinyl chloride polymers (butyl tins)are found also in animal and human livers due to diet uptake [1]. Plastic items, which can be of different polymers or shape, can be classified according to size, and in particular, items ranging within 1 μm and 5 mm are called micro plastics (MPs) [2, 3]. Micro plastics (MPs) are small pieces of plastic, less than 5 mm (0.2inch) in length, that occur in the environments resulting to plastic pollution., Physicochemical factors accelerate the degradation time of plastics by mechanical stimuli, biological, thermal, and photo-oxidative degradation [4]. Micro plastic are varied contaminant suite originated from different product types, composed of various polymers and chemical additives, characterized by a broad range of colors and shape [5].

Micro plastic consist of carbon and hydrogen atoms bond together in polymer chain, other chemicals such as phthalates, polybrominated diphenyleither's (PBDEs) and tetrabromobisphenol A (TBBPA), are present in micro plastic and many of these chemical additive leach out of the plastic after entering the river. They are divided into primary or secondary, according to their origin. Primary micro plastic (fMPs) are the final products of industrial activities used in cosmetic [6].

Primary micro plastic: include micro bends (personal care products), plastic pellets (used in manufacturing), and plastic fibers (used in synthetic textiles or nylon). Secondary micro plastics (sMPs) are products of plastics litter/degradation. Exposure of plastic waste in the environment causes deterioration in its mechanical and physicochemical properties, leading to the formation of plastic fragments, which are considered as micro plastics (MPs) when their size is less than (<5 mm). Delimi river is an important natural drainage and irrigation system, ensuring water flows through is of paramount important, as the cultivation of vegetable crops in Jos, depend on it. The presence of micro plastic in River Delimi is a health risk. The primary challenges of MPs to ecosystems are their ubiquity and bioavailability for ingestion, entanglement or inhalation [7]. MPs can be accumulated in the environment due to their inert nature [8]. They have been found in table salts and potable water [9]. The health implications of MPs may be physical by blocking the digestive system due to particle localization, chemical with associated toxic chemical effects and biological involving toxins [10]. The exposure of aquatic organisms to MPs has been associated with short- and long-term adverse effects on organism's health, including biological feeding, reproduction, antioxidant defense and innate immunity [11–14]. A contaminant that is widespread in the environment, heavy metals can enter water bodies continuously due to their non-degradable nature and are recycled and enriched in the aqueous environment. MPs and heavy metals are not only acting as persistent pollutants, their combined pollution poses a new threat to the world. Due to the large surface area, MPs can act like magnets for toxic pollutants and concentrate them to a very high level [15, 16]. The need to reduce the amount or quantity of plastic waste emitted to safeguard environmental and human health is a task to be involved every stakeholder in the environment. Heavy metals are present in the environment from both naturally occurring and anthropogenic sources. As a contaminant that is widespread in the environment, Aquatic ecosystems contain a great diversity of microorganisms, which play critical roles in many biogeochemical processes. Their existence makes the interaction between MPs and heavy metals more complicated [16]. There is no existing baseline study concerning the micro plastic challenges of River Delimi in Jos Plateau hence this study seek to provide the baseline.

1.2 The chemical composition of micro plastics

Aquatic ecosystems contain a great diversity of microorganisms, which play critical roles in many biogeochemical processes. Their existence makes the interaction between MPs and heavy metals more complicated. MPs provide an emergent ecological niche for microorganisms by the formation of microbial biofilms, named plastisphere [17, 18]. Primary micro plastics are intentionally manufactured in small sizes for different industrial purposes, such as housing and transportation applications and are generally made of polyethylene or polystyrene [19–21].

1.3 The boundaries of the study

The general aim of this study is to investigate the interaction between micro plastic and physicochemical parameters with challenges of micro plastic in River Delimi, Plateau State, Nigeria. River Delimi is within urban and commercial centre of Jos Plateau State capital, the investigated areas include; Delimi village, Delimi area, Congo Road, Tudun Wada, Anguwan Rogo, and BarakinNaraguta.

2. Materials and methods

2.1 Study area

Rive Delimi covers a distance of about 15 km as it passes through Jos metropolis. It originates from Delimi village moves through Delimi area, Anguwan Rogo, Yelwan Mista Bow, Toro Bauchi, meeting up with the Rafin Zaki and extending to the Hadeja/ Jamaare River Basin and Lake Chard. It shares a boundary to the North with Toro Local Government Area of Bauchi State; to the South with Jos-South Local Government area; to the North-East with Jos-East Local Government Area; and to the West with Bassa Local Government Area [22]. Jos enjoys a temperate climate with average temperatures of between 280C (81.70F) maximum and 110C (51.70F) minimum. It covers a total land area of 291 km^2 (112 sq mi) with a 2006 census of 429,300 people. The warmest temperatures usually occur in the dry season months of March and April Jos is characterized by a mean annual rainfall of between 1317.5 mm (131.75 cm) and 1460.00 mm (146.0 cm), mostly from May to August The Onset and Cessation of rainfall in Jos are experienced in April ($\pm$15 days in April), and October ($\pm$15 days in October). The relative humidity is characterized by a marked seasonal variation [23]. The Jos Plateau is a high land region in North Central Nigeria. River Delimi is the major drainage system in Jos Metropolis. The river serves as a source of water for domestic use, fishing and irrigation. The River Delimi in Jos Plateau is the source of many Rivers in northern Nigeria including the Kaduna Gongola, Hadejia and Yobe River [23].

Plates 1, 2 and 3, show the type of menace that the disposal of solid wastes in water resources with cumulative effect micro plastic challenges which affect the water of the urban centers and the health hazards to the people who are exposed. The waste collection containers that were seen to be available at most locations in Jos are no longer in place at most of these locations.

2.2 Data collection

2.2.1 Sampling methods and analysis

Water samples were collected from 6 different sampling points marked WT1-WT6 (plate 1–3). Bulk water sampling, which facilitates sampling of smaller micro plastics but it, is limited by the size of the area.. The collection of samples from surface water was done using bulk water technique. Temperature, electrical conductivity (EC) and total dissolved solute (TDS) were measured in-situ during the sampling process. Water samples were filtered through Millipore filter paper of reweighed 0.45 μm pore size. The filtered water samples were acidified with 0.2%(v/v) concentration $HN0_3$ and kept in glass bottles. For anions analysis the sample were not acidify and titration carried out to ascertain the concentrations of the Anions. The filter papers containing the suspended solids were air-dried and reweighd. They were digested with 4/I (v/v) HN03/HCl mixture using a microwave device. After cooling, digestions were diluted to 30 ml with mili-Q water. At each of the sampling stations three samples of plants were collected close to water sampling station while water were collected using 1 liter polyethylene bottles with screw caps which were acid washed and rinsed with distilled water prior to the sampling... A total of 16 samples comprising of different types of plants alongside water samples each from the six sites along River Delimi and River Bonga were taken for laboratory analysis (see **Tables 1** and 2). Specimens were labeled and transported to the laboratory in temperature - controlled boxes for analysis. An ambient temperature of 28° was maintained during transportation. Samples were digested with aqua regia solutions 3/I (v/v) HCl/HNO3 in microwave device. After cooling, digestions were diluted to 30 ml with mili-Q water.

2.2.2 Microplatics analysis

The different instrumental methodologies for determining and quantifying micro plastics in different matrices such as water, fruits and vegetables (see **Table 1**). The

Type	Technique (Quantification Method	Application	Size	Type of Micro plastic
Water		Bottled water and water	<100 um	PP
	Micro-Raman spectroscopy	Surface waters	0.5–5 mm	Polystyrene, polypropylene, and polyethylene
Fruits and vegetables	Electron microscopy	Lettuce (Lactuca sativa L.)	PVC-a with particle sizes from 100 nm to 18u m PVC-b with particle sizes from 18to 150 um	PVC,PET
	Scanning electron microscopy (SEM)	Cucumber plants	100, 300, 500, and 700 nm	Polystyrene nanoplastics (PSNPs)
Other	Scanning electron microscopy (SEM), X-ray	Personal care products	24–52 nm	NP polyethylene micro beads

Table 1.
Identification and quantification micro plastics in river Delimi water, fruits and vegetables.

Parameter	Sample Stations	TEM^{0}C	pH	TDS Mg/l	EC μ/m	$S0_4^{2-}$ mg/l	Cl^- mg/l	$HC0_3^-$ mg/l	Coordinates
WT1	Delimi Area	20	9.9	21	435	23.5	70.9	38.38	9°55′15" N 8° 53′48″ E
WT2	*Marhaba Masjd*	20	8.2	12.3	265	35.5	35.45	98.98	9°55′15" N 8° 53′48″ E
WT3	*Tudun Wada*	20	8.3	12.4	250	30.3	63.81	92.93	9°54′49" N 8° 53′50″ E
WT4	*Delimi Village*	20	8.2	11.1	220	30.3	28.36	88.88	9°56′01" N 8° 53′48″ E
WT5	Barkin Kogi	22	8.3	10.5	210	26.2	99.26	90.90	9^{0}56′35"N 8^0 53′04″E
WT6	Barkin Naraguta	23	8.4	10.7	213	26.5	56.72	86.86	9^{0}57′06"N 8^{0}52′45″E

Table 2.
Concentration of physical parameters and anions in water from all the sampling sites.

techniques are focused on a physical type (non-destructive) determination; the composition of the micro plastic of the polymer that was separated from the sample by different analytical procedures is visualized. Reagents for separating the micro plastics from the samples [3, 19, 24, 25] with a high concentration of organic matter, where their action is to destroy the organic matter and release the micro plastics from the matrix in order to later be separated by filtration processes [7, 26–30]. Mesoplastic (MEP,>5 mm) and coarse micro plastic (Cmp,>2 mm) were visually identified from sieving fractions using a stainless-steel bowl with an imprinted grid (1x1 cm grid size) and inspected under stereomicroscope. Large micro plastics (L-Mp,>1 mm) and medium micro plastics (M-Mp,>0.3 mm) were analyzed after density separation micro plastics ZnCl2 and NaI. Separating micro plastics from water samples was done either by direct filtration [7, 31, 32] or organic materials such as dried leaves which showed recovery rate of 82% [33]. For density separation, solution of chemicals such as NaCl, NaI, NaCl, and ZnCl2 are used. After density separation, sieving, and filtration of floating sample material [30, 34]. Stained filters were visually detected under a stereomicroscope. Concentrated samples retained on the filters were identified and quantified. In the identification and quantification of micro plastics in samples, light microscopes have been applied in studies conducted in laboratories. Final identification of each potential plastics particle within the MEP, CMp, and L-Mp and M-Mp sizes was performed.

3. Results and discussion

3.1 Concentration of physical parameters (tem, pH, TDS and EC)

Figure 1 illustrates the concentration of Physical parameters in all sample locations. The highest concentrations of electrical conductivity (EC) are in sample locations WT1 (near Delimi street Bridge) along Delimi river and WT2 (katako) also along the Delimi river. Concentrations were lower in WT5-WT6 (AngwanRogo) (Naraguta) and along the Delimi, both along River Bonga.

The characteristic of physical parameters of samples from six samples that were analyzed for physical parameters in River Delimi (see **Table 2**).

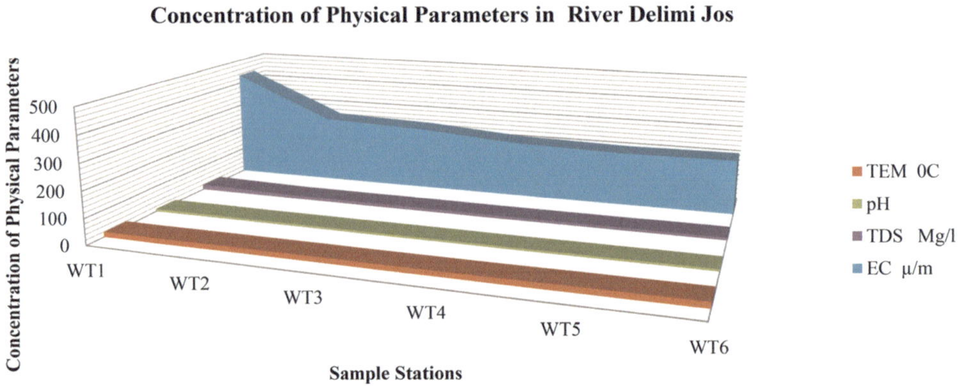

Figure 1.
Concentration of physical parameters in water from the entire sampling site.

Variation of physical parameters in River Delimi showed EC has highest concentration at WT1-WT2. Value of electrical conductivity EC of the water sample at WT1 (435 ohm/m), while WT5 exhibits lowest value of EC (210 ohm/m). Temperatures, pH and TDS have lower values (presented in **Figure 1**).

3.2 Concentration of anions ($S04^{2-}$, $Cl^{-}$and $HC03^{-}$)

The concentration of Anions in water from all sample location. The concentration of hydrocarbon ate was high in sample WT2 –WT6 (Congo) along river Bonga and chloride at sample and WT5 –WT6 (Anguwan Soya) along river Bonga. However, the lowest concentration was recorded at sample station WT1 (Delimi street Bridge) for hydrocarbon ate and WT4 (Anguwan Rogo) for chloride.

The highest (see **Table 2** and presented in **Figure 2**) concentration of hydrocarbonate 98.98 (WT2) while the lowest concentration 38.38 (WT1), chloride highest concentration 99.26 (WT5) while the lowest concentration 28.36 (WT4) and sulfite which the least,highest concentration 35.5 (WT2) while lowest concentrations 23.5 (WT1). Hydrocarbon ate and chloride has higher values in River Delimi.

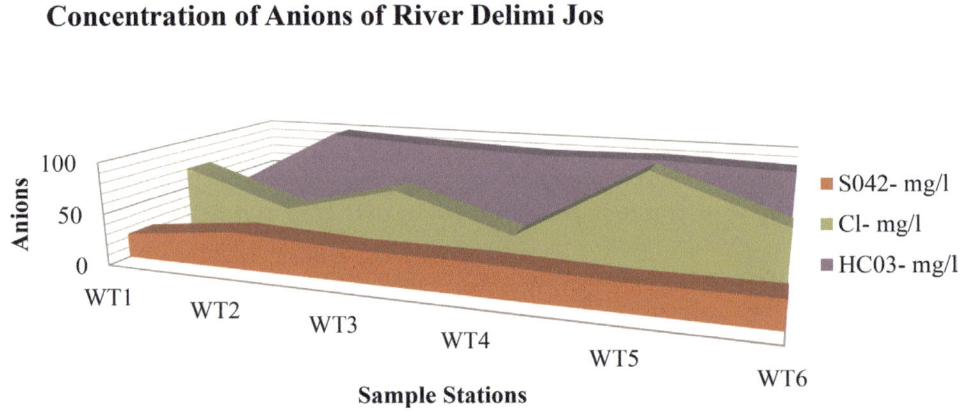

Figure 2.
Concentration of anions in water from the entire sampling site.

Parameter	Sample Stations	Na^+ mg/l	K^+ Mg/l	Mg^{2+} Mg/l	Ca^{2+} Mg/l	Cr Mg/l	Ni Mg/l	Pb Mg/l	Fe Mg/l	Zn Mg/l	Coordinates
WT1	Delimi area	17.38	4.43	0.49	14.43	3.62	0	2.36	0	0.12	9°55′15" N 8° 53′48″ E
WT2	*Marhaba Masja*	18.04	4.44	0.48	14.50	3.55	0	2.28	0.21	0.08	9°55′06" N 8° 53′48″ E
WT3	*Tudun Wada*	18.50	4.48	0.47	14.49	3.17	0	2.29	0	0.09	9°54′49" N 8° 53′50″ E
WT4	*Delimi Village*	17.20	4.50	0.46	14.48	2.91	0	2.30	0	0.09	9° 51 '30" N 8° 56′01″ E
WT5	Barakin Kogi	16.71	4.54	0.45	14.43	2.83	0	2.31	0	0.1	9^0 56′35"N $8^0$53′04″E
WT6	Barakin Naraguta	14.48	4.57	0.40	14.42	2.77	0	2.32	0	0.1	9^0 57′06"N 8^0 52′45″E

Table 3.
Concentration of cations and heavy metals in water from all the sampling sites.

3.3 Concentration of cations (Na^+, K^+, Mg^{2+} and Ca2+)

The concentration of cations in all sample locations. The concentration of **Sodium (Na^+) and calcium (Ca^{2+})** are higher in the sample locations from WT1-WT6 both along river Delimi and Bonga. Highest (see **Table 3** and presented in **Figure 3**) concentration of sodium in river Delimi 18.50 (ST6) mg/l while the lowest 14.48 (WT6), calcium highest concentration 14.50 (WT2) while the lowest concentration 14.32 (WT6).

3.4 Concentration of heavy metals (Cr, Ni, Pb and Fe)

The concentration of heavy metals in all sample locations. The concentration of chromium is higher in all the sample locations from WT1-WT6 both along river Delimi and Bonga. Lead is higher in the sample locations WT4-WT6. **Figure 4**

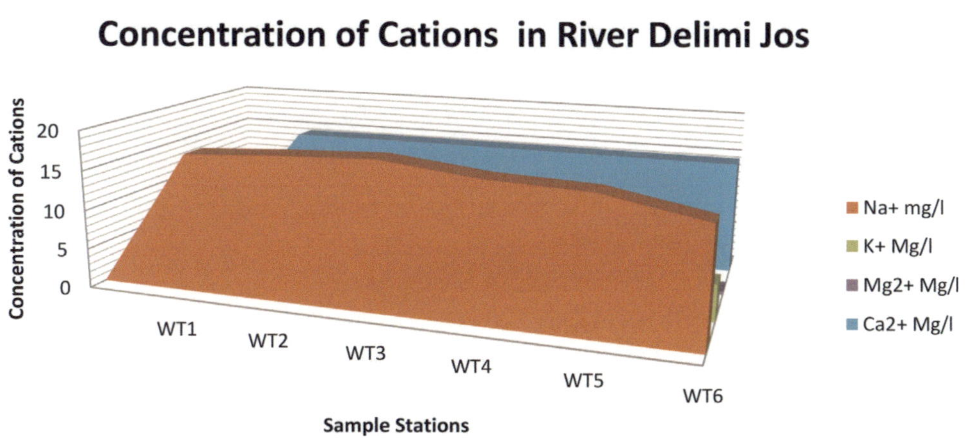

Figure 3.
Concentration of cations in water from the entire sampling site.

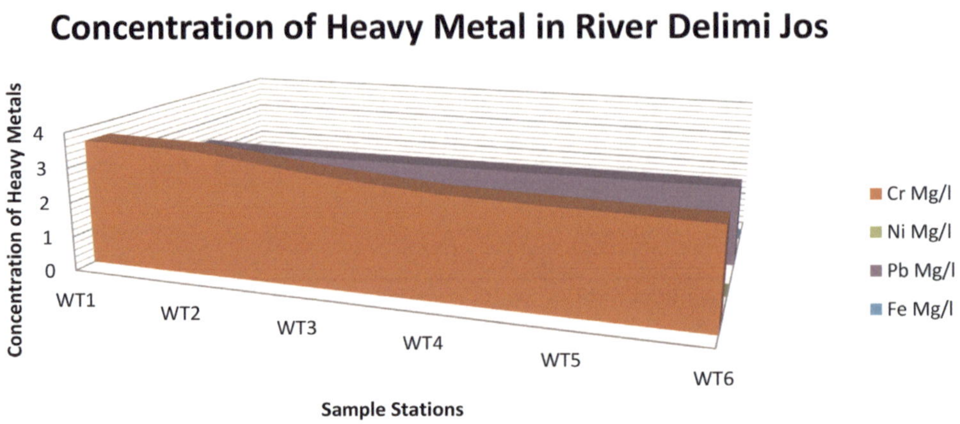

Figure 4.
Concentration of heavy metals in water from the entire sampling site.

revealed high concentration of chromium 3.62 (WT1) and lowest value of 2.77 (WT6) while lead higher value 2.36 (WT1), least value 2.28(WT2) (see **Table 3** and presented in **Figure 4**).

Micro plastics in River Delimi.

Concentration o f micro plastics (MPs) in river Delimi both dry and rainy season were abundant in color and dominated by fibrous items. Polyethylene terephthalate (PET) concentration 22.65% and Plasticized polyvinylchloride (PVC) concentration 50% while others 27.55% (dry season). Polyethylene terephthalate (PET) concentration 18.20%(and Plasticized polyvinylchloride (Plasticized (PVC) concentration 80.50% while others 11.70% (rainy season) Plasticized polyvinylchloride (Plasticized (PVC) were the predominant accounting for 50% and 80.50% in both dry and rainy season (presented in **Figures 5** and **6**).

3.5 Concentration of metals (Ca^{2+}, Mg^{2+} and K^{+})

The concentration of potassium (K^{+}) in all sample locations. The concentration of potassium is higher in all the sample locations from WT1-WT16 both along river Delimi and Bonga (see **Tables 4** and **5** and presented in **Figure 7**).

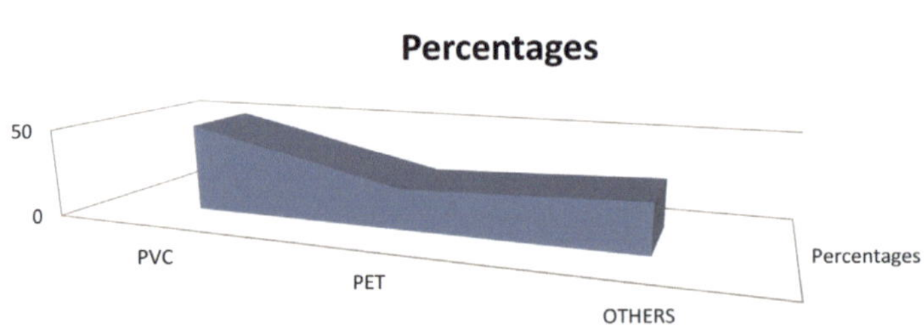

	PVC	PET	OTHERS
Percentages	50	22.65	27.35

Figure 5.
Concentration of micro plastic in river Delimi water from the entire sampling site (dry season).

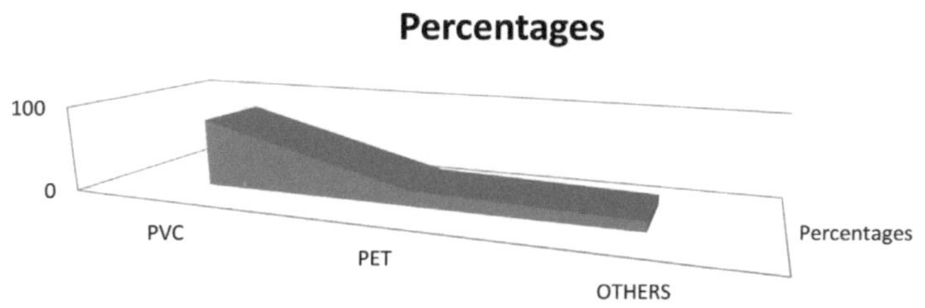

	PVC	PET	OTHERS
■ Percentages	80.5	18.2	11.7

Figure 6.
Concentration of micro plastics in river Delimi water from the entire sampling site (rainy season).

3.6 Concentration of nonmetals (S and P)

The concentration of sculpture and phosphorus in PT4-PT16 sample locations. The concentration of sulfur and phosphorus in sample locations PT4-PT16 are higher both along river Delimi and Bonga (presented in **Figure 8**).

3.7 Concentration of metals and nonmetals

The concentration of metals and nonmetals in all sample locations. The concentration of metals is higher in all the sample locations from PT1-PT16 both along river Delimi and Bonga (Presented in **Figure 9**).

3.8 Concentration of heavy metals (Mo, Cu, Pb and Zn)

The concentration of heavy metals n all sample locations. The concentration of molybdenum is higher in all the sample locations from PT1-PT16 both along river Delimi and Bonga (see **Table 6** and presented in **Figure 10**).

3.9 Concentration of heavy metals (Ni, Co, Mn, Cr, Ba, Ti and B)

The concentration of heavy metals in all sample locations. The concentration of Mn is higher in all the sample locations from W T1-WT16 both along river Delimi and Bonga. The concentration of B, Ti and Ba also show high concentration (presented in **Figure 11**).

3.10 Concentration of heavy metals (W, Sc, Tl, Se and Ga)

The concentration of metals in all sample locations. The concentration of metals that are less than or equal to one in all the sample locations from WT1-WT16 both along river Delimi and Bonga..Ba, Ce, Rb, La, Nd, Ta, Sm, Sc and Th were high (presented in **Figure 12**).

3.11 Interaction between MPs and heavy metals in plants irrigated with river Delimi water

Micro plastics particles can interact with biological and synthetic systems due to their small size, enormous specific surface area, and high functionalization capacity,

PT1 PT2 PT3	**Tomatoes**	**Delimi Area along Rive Delimi**	**Solanum lycopersicum**	**9°55′15” N 8° 53′48″ E**
	Spinach	**Delimi Area along River Delimi**	**Spinacia oleracea**	**9°55′15” N 8 ° 53′48″ E**
	Banana	**Delimi Area along River Delimi**	**Musa acuminata**	**9°55′15” N 8 ° 53′48″ E**
PT4 PT5 PT6	Pear	Marhaba Masjd along River Delimi	Pyrus communis	9°55′15” N 8 ° 53′48″ E
	Mango	Marahaba Masjd along River Delimi	Mangifera indica	9°55′15” N 8° 53′48″ E
	Tomatoes	Marhaba Masjd along River Delimi	S. lycopersicum	9°55′15” N 8° 53′48″ E
PT7 PT8 PT9	Cassava	Tudun Wada along River Delimi	Manihot esculenta	9°54′49” N 8° 53′50″ E
	Tomatoes	Tudun Wada along River Delimi	S. lycopersicum	9°54′49” N 8° 53′50″ E
	Tomatoes	Tudun Wada along River Delimi	Solanum lycopersicum	9°54′49” N 8° 53′50″ E
PT10 PT11 PT12	Tomatoes	Malam Adam along River Delimi	Solanum Iycopersicum	9° 55 ‘45” N 8° 53′39″ E
	Sweet Pepper	Malam Adam along River Delimi	Capsicum annuum	9° 55 ‘45” N 8° 53′39″ E
	Shambo Pepper	Malam Adam along River Delimi	C. annuum	9° 55 ‘45” N 8° 53′39″ E
PT13 PT14 PT15	Sweet Pepper	Barakin Kogi along River Bunga	C. annuum	9^{0} 56′35”N 8 ^{0}53′04″E
	Shambo Pepper	Barakin Kogi along River Bunga	C. annuum	9^{0} 56′35”N 8 ^{0}53′04″E
	Tomatoes	Barakin Kogi along River Bunga	Solanum Iycopersicum	9^{0} 56′35”N 8 ^{0}53′04″E
PT16 PT17 PT18	Irish Potatoes	Irish Potatoes	Solanum tuberosum	9^{0} 57′06”N 8^{0} 52′45″E
	Cucumber	Cucumber	Cucumis sativus	9^{0} 57′06”N 8^{0} 52′45″E
	Tomatoes	Tomatoes	S. lycopersicum	9^{0} 57′06”N 8^{0} 52′45″E

Table 4.
Concentration of metals and non metals in plants irrigated with river Delimi water from all the sampling sites.

capable of even permeating biological membranes [3] Various sources of water and soils contaminated with micro plastics, are the source from where plants absorb water and nutrients.

The consumption of vegetable crops with a high level of heavy metals (Ba, Ce, Rb, La, Nd, Ta, Sm, Sc and Th) while mercury and silver show high concentration (Hg, Ag) (see **Tables 6** and 7) due to high content of PVC in both dry and rainy season which is a challenge to humans. This showed that the use of water from river Delimi for irrigation may cause danger to the crops growing. Micro plastics increase cations of

Method	1VE	1VE	1VE	1VE	1VE	1VE	1VE
Analyte	Ca	Mg	K	Na	Al	P	S
Unit	%	%	%	%	%	%	%
MDL	0.01	0.001	0.01	0.001	0.01	0.01	0.01
PT1	0.2	0.224	5.14	0.283	0	0.72	0.25
PT2	3.08	0.762	6.53	0.017	0.1	0.34	0.34
PT3	0.03	0.15	1.63	<0.001	<0.01	0.16	0.1
PT4	0.09	0.097	2.33	0.002	<0.01	0.22	0.16
PT5	0.16	0.081	1.24	0.005	<0.01	0.14	0.12
PT6	0.26	0.316	6.05	0.403	0	0.58	0.33
PT7	0.06	0.063	1.28	0.002	<0.01	0.25	0.07
PT8	0.17	0.263	5.79	0.188	<0.01	0.75	0.34
PT9	0.26	0.224	4.71	0.036	<0.01	0.57	0.25
PT10	0.29	0.213	2.8	0.014	<0.01	0.51	0.41
PT11	0.1	0.176	2.98	0.018	<0.01	0.41	0.27
PT12	0.08	0.194	3.97	0.032	<0.01	0.4	0.2
PT13	0.13	0.263	6.22	0.173	0	0.57	0.29
PT14	0.04	0.13	2.45	0.035	<0.01	0.31	0.2
PT15	0.45	0.272	5.2	0.040	<0.01	1.02	0.27

Table 5.
Concentration of heavy metals in plants irrigated with river Delimi water from all the sampling sites.

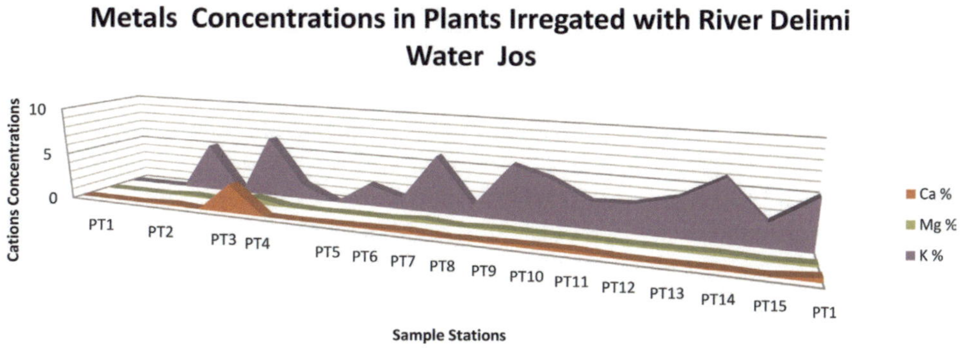

Figure 7.
Concentration of metals in plants irrigated with river Delimi water from the entire sampling site.

Na^+, Ca^{2+} and decrease K^+, Mg^{2+}, and Fe concentrations. Fragment, fiber shapes were identified in the surface water of the rivers, with fragment shape having the highest occurrence. The distribution of the plastics was as follows: polyethylene terephthalate (PET) – 22.65%, polyvinyl chloride (PVC) – 50%, others – 27.35% (Dry season), (PET) – 18.20%, polyvinyl chloride (PVC) – 80,50%, others – 11.70% (Rainy season) [35]. Estimated that humans are exposed to about 27 types of micro pollutants by consuming fruits and vegetables derived from irrigation water contaminated with

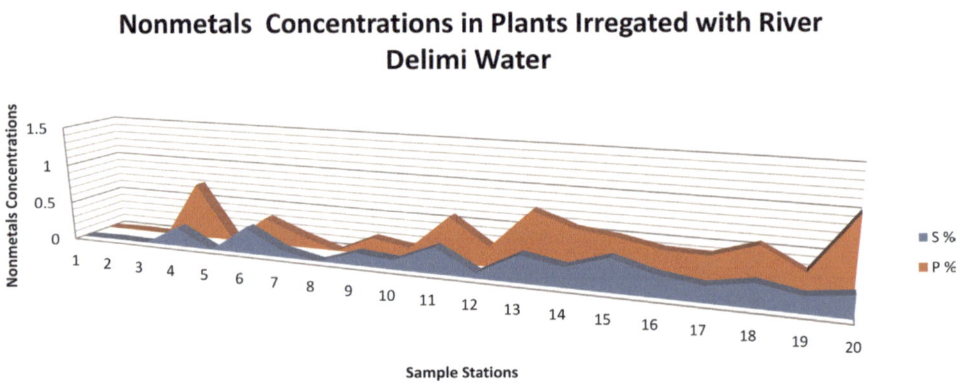

Figure 8.
Concentration of nonmetals in plants irrigated with river Delimi water from the entire sampling site.

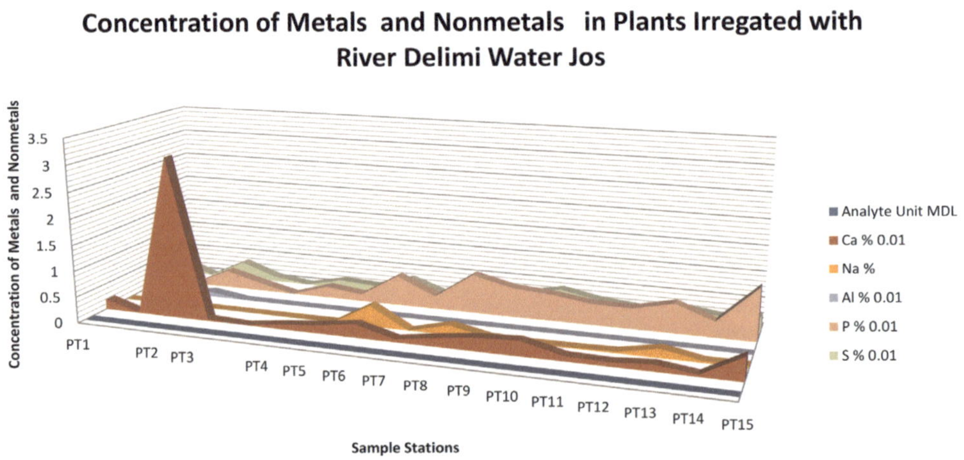

Figure 9.
Concentration of metals and nonmetals in plants irrigated with river Delimi water from the entire sampling site.

these compounds [36], indicates that the concentration of micro plastics. Environmental factors may indirectly exert influence on MPs by changing biofilm structures on the MPs surface [10] found that nutrient salts, total nitrogen, total phosphorus, and pH have a greater influence on colony structure, while MP physical and chemical properties such as particle size and contact angle have less influence. Both MPs and heavy metals can accumulate at high level in the environment and consequently contaminate the food. Many studies have shown that MPs can absorb and release heavy metals, and their combined exposure may pose a potential threat to ecological system and human being [37].

3.12 Effects of environmental factors on interactions between MPs and physicochemical parameters

The result of physic-chemical properties has been evaluated with a statistical test. The result computed between physicochemical properties as it affect the concentrations of different micro plastic formed in River Delimi presented in **Figures 1–4**. The

Method	1VE	1VE	1VE	1VE	1VE	1VE	1VE	1VE	1VE
Analyte	Mo	Cu	Pb	Zn	Ag	Ni	Co	Mn	Fe
Unit	ppm	ppm	ppm	ppm	ppb	ppm	ppm	ppm	%
MDL	0.01	0.01	0.01	0.1	2	0.1	0	1	0.001
PT1	1.02	12.49	0.76	33	4	1.1	0.2	14	0.009
PT2	1.71	9.5	4.45	89	19	1.4	0.5	48	0.08
PT3	0.54	6.65	0.03	8.5	<2	0.3	<0.01	5	0.003
PT4	0.18	11.71	0.06	29	<2	1.2	0.4	4	0.005
PT5	0.04	2.06	0.12	11	<2	0.3	0	17	0.005
PT6	1.29	14.52	0.27	101	8	0.5	0.2	19	0.013
PT7	0.02	4.71	0.38	22	<2	2.5	0.3	8	0.003
PT8	1.58	12.56	0.24	40	7	0.8	0.1	15	0.012
PT9	0.76	17.55	0.14	40	4	1.1	0.1	19	0.01
PT10	0.26	17.76	0.12	47	4	2.7	0.2	44	0.011
PT11	0.38	11.87	0.06	22	<2	1.6	0.4	20	0.006
PT12	0.37	10.97	0.08	31	5	1.1	0.2	23	0.008
PT13	2.32	15.37	0.15	33.3	6	0.4	0.11	20	0.012
PT14	0.11	10.6	0.1	26	4	0.4	0.1	8	0.005
PT15	3.41	8.5	0.18	32	5	1.1	0.07	6	0.009

Table 6.
Concentration of heavy metals in plants irrigated with river Delimi water from all the sampling sites.

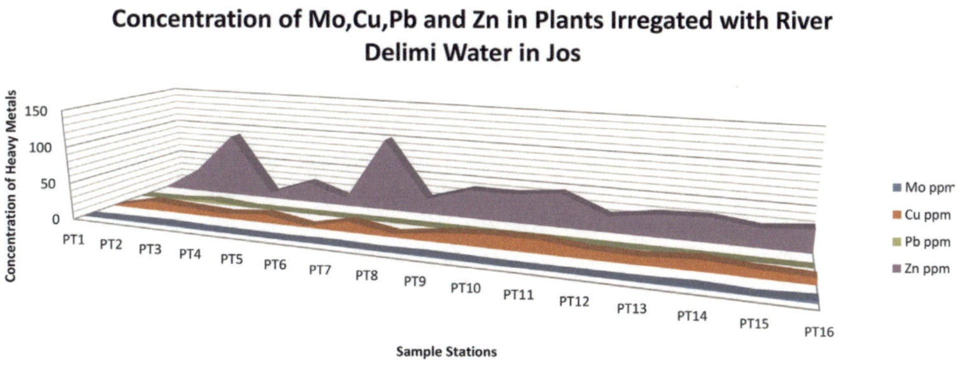

Figure 10.
Concentration of heavy metals in plants irrigated with river Delimi water from the entire sampling site.

concentration of physical parameters were statistically tested. The result showed different environmental factors interactions between physicochemical parameters and micro plastics(MPs). This study considered the following factors:. Temperature, pH, electrical conductivity (EC) contact time, ionic strength, and particle size. As for the temperature, in this study the temperature range from 20 to 23°C (**Figure 3**), the general opinion is that high temperature will benefit the adsorption of heavy metals

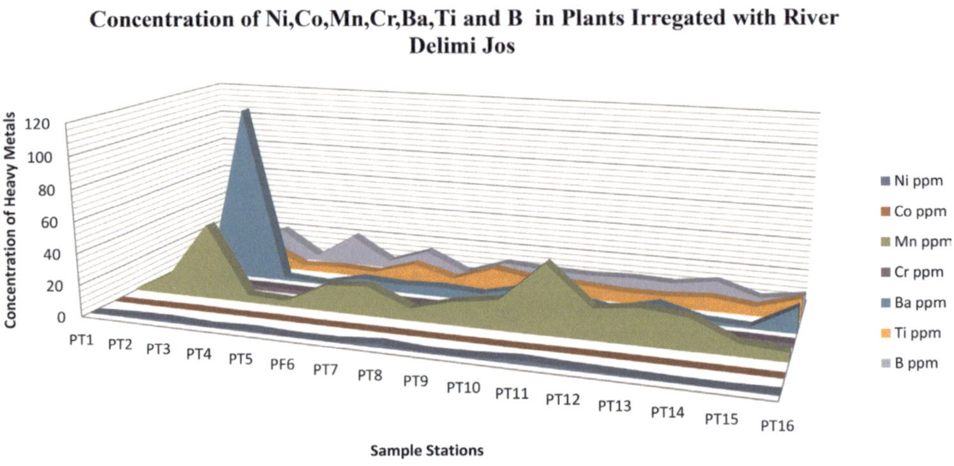

Figure 11.
Concentration of heavy metals in plants irrigated with river Delimi water from the entire sampling site.

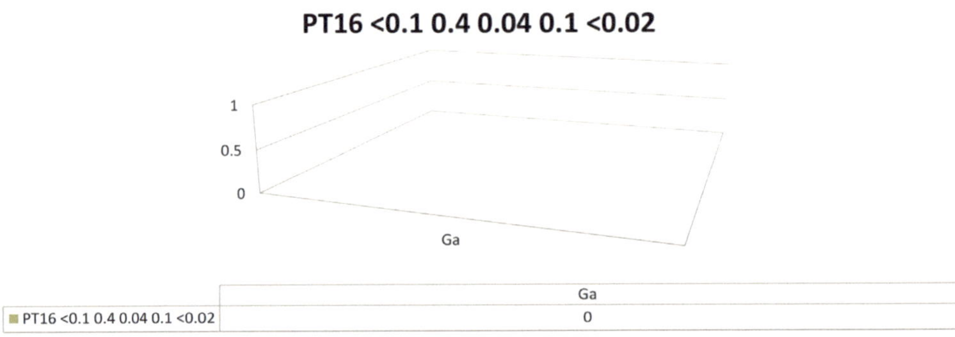

Figure 12.
Concentration of Heavy Metals in Plants Irrigated with River Delimi Water from the Entire Sampling Site.

on MPs [10, 38]. The possible explanation for this observation is that the adsorption process is an endothermic reaction; thus, the spontaneity of the adsorption process may increase with the increase of temperature.The pH, in the study area range from 8.2 to 9.9.The pH is greater than 7.(pH < 7) and metal ions were considered to precipitate under alkaline conditions. The pH can significantly affect the adsorption capacity of MPs to heavy metals. pH is very important for mobility because metal availability is low when pH, is around 6.5 to 7, MPs will not interact with heavy metals. Generally, increased pH level results in increased adsorption capacity for heavy metals. The absorbed heavy metals released with the change of pH in water (see **Tables 1** and **2**).

3.13 Challenge of Microplastis to human health

The increasing amount of heavy metals has caused an imbalance in aquatic ecosystems and the biota growing under such habitats accumulate high amounts of heavy metals (Cu, Zn, Cd, Cr, and Ni) which are being assimilated and transferred within food chains by the process of magnification. Many researchers tried to use different

Method	**1VE**	**1VE**	**1VE**	**1VE**	**1VE**	**1VE**	**1VE**	**1VE**	**1VE**	**1VE**	**1VE**
Analyte	Cr	Ba	Ti	B	W	Sc	Tl	Hg	Se	Te	Ga
Unit	ppm	ppm	ppm	ppm	ppm	ppm	ppm	ppb	ppm	ppm	ppm
MDL	0.1	0.1	1	1	0.1	0.1	0.02	1	0.1	0.02	0.1
PT1	2.6	7.6	14	10	<0.1	0.3	<0.02	95	0.1	<0.02	<0.1
PT2	2.8	113.6	16	20	<0.1	0.3	0.17	162	0.3	0.03	0.3
PT3	3	2.5	4	4	<0.1	0.5	<0.02	24	0.2	<0.02	<0.1
PT4	2.6	1	5	20	<0.1	0.3	<0.02	55	0.2	<0.02	<0.1
PT5	2.8	6.8	4	5	<0.1	0.4	<0.02	55	0.2	<0.02	<0.1
PT6	2.9	5.5	12	14	<0.1	0.3	<0.02	115	0.1	<0.02	<0.1
PT7	3	6.4	5	2	<0.1	0.3	<0.02	23	0.2	<0.02	<0.1
PT8	2.9	4.9	14	11	<0.1	0.3	<0.02	135	0.2	<0.02	<0.1
PT9	3.2	11.1	10	9	<0.1	0.2	<0.02	125	0.2	<0.02	<0.1
PT10	2.9	8.2	9	8	<0.1	0.2	0.4	220	0.1	<0.02	<0.1
PT11	3.3	3.3	7	9	<0.1	0.4	0.42	156	0.1	<0.02	<0.1
PT12	3.1	9.7	7	8	<0.1	0.4	<0.02	116	0.2	<0.02	<0.1
PT13	2.9	2.5	9	12	<0.1	0.2	<0.02	87	0.2	<0.02	0.1
PT14	3.1	1.3	6	4	<0.1	0.5	0.06	35	0.2	<0.02	<0.1
PT15	2.6	16.3	14	8	<0.1	0.4	0.04	116	0.1	<0.02	<0.1

Table 7.
Concentration of heavy metals in plants irrigated with river Delimi water from all the sampling sites.

test methods or models to assess the potential human health risks of MPs. The results showed that physicochemical parameters concentration in River Delimi undergo changes due to micro plastics concentration.

River Delimi water has moderately interacted with microplastics. Physicochemical parameters especially heavy metals,bond to the microplastic. Emphasis should be placed on protection of river Delimi considering its sensitive nature of its use for irrigation and addition of value to food chain, toward reducing micro plastic interaction with physicochemical parameters.

4. Conclusion

The observed interaction between physicochemical parameters and micro plastics in Delimi River is a general one affecting the majority of inland waters of Nigeria. It is evident that, there is a serious problem with micro plastic interaction in River Delimi and its environs. The reason is mainly as a result of population increase in the Jos metropolis leading to enormous solid wastes generation. The use and throw away attitude of the residents have contributed immensely in the quantity of plastic generated every day. Physicochemical analysis and interaction with micro plastic items can contribute to the water pollution: that is used for irrigation of crops planted within the proximity of River Delimi.

Author details

Terwase Wuave and Ahmed Sabo*
Department of Environmental Management Technology, Abubakar Tafawa Balewa University Bauchi and National Environmental Standards Regulations and Enforcement Agency (NESREA), Lushi, Nigeria

*Address all correspondence to: asabo@atbu.edu.ng

References

[1] Takahashi S, Mukai H, Tanabe S, Sakayama K, Miyazaki T, Masuno H. Butyltin residues in livers ofhumans and wild terrestrial mammals and in plastic products. Environmental Pollution. 1999;**106**:213-218

[2] Thompson RC, Moore CJ, vom Saal FS, Swan SH. Plastics, the environment and human health: Currentconsensus and future trends. Philosophical Transactions of the Royal Society B. 2009;**364**:2153-2166

[3] Gigault J, Halle AT, Baudrimont M, Pascal P-Y, Gau_re F, Phi T-L, et al. Current opinion: What is a nanoplastic? Environmental Pollution. 2018;**235**: 1030-1034

[4] Singh B, Sharma N. Mechanistic implications of plastic degradation. Polymer Degradation and Stability. 2008;**93**:561-584

[5] Rochman CM, Brookson C, Bikker J, Djuric N, Earn A, Bucci K, et al. Et a rethinking microplastics as a diverse contaminant suite. Environmental Toxicology and Chemistry. 2019;**38**: 703-711

[6] Guerranti C, Martellini T, Perra G, Scopetani C, Cincinelli A. Microplastics in cosmetics: Environmental issues and needs for global bans. Environmental Toxicology and Pharmacology. 2019;**68**:75-79

[7] Enyoh CE, Shafea L, Verla W, Verla N, Wang Q, Chowdhury T, et al. Microplastics exposure routes and toxicity studies to ecosystems: An overview. Environmental Analysis Health and Toxicology. 2020;**35**: e2020004. DOI: 10.5620/eaht.e2020004

[8] Wang J, Li J, Liu S, Li H, Chen X, Peng C, et al. Distinct microplastic distributions in soil of different land-use types: A case study of Chinese farmlands. Environmental Pollution. 2021;**269**:11619. DOI: 10.1016/j.envpol.2020.116199

[9] Verla AW, Enyoh CE, Verla EN, Nwarnorh KO. Micro plastic toxic chemical interaction: A review study on quantified levels mechanism and implication. SN Applied Science. 2019;**1**:1400. DOI: 10.1007/s42452-019-1352-0

[10] Wang Q, Zhang Y, Wangjin X, Wang Y, Meng G, Chen Y. The adsorption behavior of metals in aqueous solution by microplastics effected by UV radiation. Journal of Environmental Sciences. 2020;**87**:272-280. DOI: 10.1016/j.jes.2019.07.006

[11] Murphy F, Quinn B. The effects of microplastic on freshwater Hydra attenuata feeding, morphology & reproduction. Environmental Pollution. 2018;**234**:487-494. DOI: 10.1016/j.envpol.2017.11.029

[12] Wen B, Jin S, Chen Z, Gao J, Liu Y, Liu J, et al. Single and combined effects of microplastics and cadmium on the cadmium accumulation, antioxidant defence and innate immunity of the discus fish (Symphysodon aequifasciatus). Environmental Pollution. 2018a;**243**:462-471. DOI: 10.1016/j.envpol.2018.09.029

[13] Banaee M, Soltanian S, Sureda A, Gholamhosseini A, Haghi BN, Akhlaghi M, et al. Evaluation of single and combined effects of cadmium and micro-plastic particles on biochemical and immunological parameters of common carp (Cyprinus carpio). Chemosphere. 2019;**236**:124335. DOI: 10.1016/j.chemosphere.2019.07.066

[14] Oliviero M, Tato T, Schiavo S, Fernandez V, Manzo S, Beiras R. Leachates of micronized plastic toys provoke embryotoxic effects upon sea urchin Paracentrotus lividus. Environmental Pollution. 2019;**247**: 706-715. DOI: 10.1016/j.envpol.2019.01.098

[15] Liu S, Shi J, Wang J, Dai Y, Li H, Li J, et al. Interactions between microplastics and heavy metals in aquatic environments: A review. Frontiers in Microbiology. 2021;**2021**:652520. DOI: 10.3389/fmicb.2021.652520

[16] Prunier J, Maurice L, Perez E, Gigault J, Pierson-Wickmann A, Davranche M, et al. Trace metals in polyethylene debris from the North Atlantic subtropical gyre. Environmental Pollution. 2019;**245**:371-379. DOI: 10.1016/j.envpol.2018.10.043

[17] Mincer TJ, Zettler ER, Amaral-Zettler LA. Biofilms on plastic debris and their influence on marine nutrient cycling, productivity, and hazardous chemical mobility. Handbook of Environmental Chemistry. 2016;**78**:221-233. DOI: 10.1007/698_2016_12

[18] Yang Y, Liu W, Zhang Z, Grossart H, Gadd GM. Microplastics provide new microbial niches in aquatic environments. Applied Microbiology and Biotechnology. 2020;**104**:6501-6511. DOI: 10.1007/s00253-020-10704-x

[19] Hernandez LM, Yousefi N, Tufenkji N. Are there Nanoplastics in your personal care products? Environmental Science & Technology Letters. 2017;**4**:280-285

[20] Mattsson K, Hansson L-A, Cedervall T. Nano-plastics in the aquatic environment. Environmental Science. Processes & Impacts. 2015;**1**:1712-1721

[21] Huang Y, Chapman J, Deng Y, Cozzolino D. Rapid measurement of microplastic contamination in chicken meat by mid infrared spectroscopy and chemometrics: A feasibility study. Food Control. 2020;**113**:107187

[22] Mailoushi JT, Binbol NL, Sadiku Y. Measure of evaporation in Jos. A comparative analysis. In: Iliya MA, Adamu IA, Umar AT, Audu M, editors. Climate Change, Environmental Challenges and Sustainable Development. International Conference on Climate Change, Environmental Challengesand Sustainable Development. 23rd – 26th November. Sokoto: Usmanu Danfodiyo University; 2015. pp. 25-29

[23] Sabo A, Nayaya AJ, Galadima AI. Assessment of some heavy metals in water, sediment and freshwater mudfish. 2008

[24] Blair RM, Waldron S, Phoenix V, Gauchotte-Lindsay C. Micro- and Nanoplastic Pollution of Freshwater and Wastewater Treatment. Vol. 5. Systems Springer Science Reviews; 2017. pp. 19-30

[25] Puoci F, Iemma F, Spizzirri UG, Cirillo G, Curcio M, Picci N. Polymer in agriculture: A review. American Journal of Agricultural and Biological Sciences. 2008;**3**:299-314

[26] Dehaut A, Cassone A-L, Frère L, Hermabessiere L, Himber C, Rinnert E, et al. Microplastics in seafood: Benchmark protocol for their extraction and characterization. Environmental Pollution. 2016;**215**:223-233

[27] Phuong NN, Poirier L, Lagarde F, Kamari A, Zalouk-Vergnoux A. Microplastic abundance and characteristics in French Atlantic coastal sediments using anew extraction

method. Environmental Pollution. 2018; **243**:228-237

[28] Simon M, van Alst N, Vollertsen J. Quantification of microplastic mass and removal rates at wastewater treatment plantsapplying focal plane Array (FPA)-based Fourier transform infrared (FT-IR) imaging. Water Research. 2018;**142**: 1-9

[29] Nuelle M-T, Dekiff JH, Remy D, Fries E. A new analytical approach for monitoring microplastics in marine sediments. Environmental Pollution. 2014;**184**:161-169

[30] Gao M, Xu Y, Liu Y, Wang S, Wang C, Dong Y, et al. Effect of polystyrene on di-butyl phthalate (DBP) bioavailability and DBP-induced phytotoxicity in lettuce. Environmental Pollution. 2021;**268**:115870

[31] Abiodun OA, Osuala FI, Otitoloju AA, Fotsing CMD, Ndinteh DT. Micropellet particles: A vector of hydrophobic endocrine disturbing chemicals in Lagos lagoon. Current Journal of Applied Science and Technology. 2019;**36**:1. DOI: 10.9734/cjast/2019/v36i630262

[32] Olarinmoye OM, Stock Scherf F, Whenu OO, Asenime CE, Ganzallo S. Geosciences. 2020;**10**:494. DOI: 10.3390/geosciences10120494

[33] Briggs E, de Moura EAB, Furusawa HA, Cotrim MEB, Oguzie EE, Lugao AB. The minerals, metals and materials series. In: Li B et al., editors. Characterization of Minerals, Metals, and Materials. 2019. DOI: 10.1007/978-3-030-05749-7_27

[34] Thompson RC, Olsen Y, Mitchell RP, Davis A, Rowland SJ, John AWG, et al. Lost at sea: Where is all the plastic? Science. 2004;**304**:838

[35] Hermabessiere L, Dehaut A, Paul-Pont I, Lacroix C, Jezequel R, Soudant P, et al. Occurrence and effects of plastic additives on marine environments and organisms: A review. Chemosphere. 2017;**182**:781-793

[36] Akhbarizadeh R, Moore F, Keshavarzi B. Investigating a probable relationship between microplastics and potentially toxicelements in fish muscles from northeast of Persian gulf. Environmental Pollution. 2018;**232**: 154-163

[37] Akhbarizadeh R, Moore F, Keshavarzi B, Moeinpour A. Microplastics and potentially toxic elements in coastal sediments of Iran's main oil terminal (Khark Islan D). Environmental Pollution. 2017;**220**: 720-731. DOI: 10.1016/j.envpol.2016.10.038

[38] Oz N, Kadizade G, Yurtsever M. Investigation of heavy metal adsorption on microplastics. Applied Ecology and Environmental Research. 2019;**17**: 7301-7310. DOI: 10.15666/aeer/1704_73017310

Chapter 8

Microplastics Derived from Commercial Fishing Activities

Tore Syversen and Grethe Lilleng

Abstract

Ordinary fishing activity is a source of microplastics to the sea that is often overlooked and scarcely reported in the literature. In this paper, we estimate the number of microplastics in the ocean that originates from the wear and tear of different fishing gear used during ordinary, commercial fishing. The wear comes mainly from rope abrasion caused by the haulers and gear dragged along the sea bottom. The types of fishing gear considered are pots, gillnets, longlines, Danish seine, and trawls. Our calculations show that about 208 tons of microplastics are produced annually from the Norwegian fishery. Globally, it sums to 4 622 tons annually. However, the calculations have several questionable parameters, and these numbers must be considered a first rough estimate of the generated microplastics. More research is needed to get better estimates, particularly regarding trawl dolly ropes.

Keywords: microplastics, wear and tear, gillnet, crab pots, longlines, Danish seine, trawls, fishing ropes

1. Introduction

Plastic pollution in the sea is a widespread problem that has gained much focus recently. One of the most prominent sources of this pollution is fishing gear accounting for about 18% of the total marine plastic debris [1]. This plastic pollution causes lots of damage to the wildlife in and around the oceans. Some review papers discussing this damage are [2–8]. Furthermore, several studies have reported on the occurrence of microplastics in marine animals from the Middle East [9], Europe [10–15], Asia [16], South America [17], Africa [18], and Australia [19].

This chapter reports the plastic pollution caused by fishing gear during ordinary fishing activities, that is, wear and tear from the plastic ropes due to sea bottom contact and abrasion caused by the hauling equipment. Modern fishing gear is composed of different plastics, with polyethene (PE), polypropylene (PP), and polyamide (PA) being the most widely used [1]. These plastic ropes are worn during everyday use and regularly replaced when the wear gets too high. The amount of microplastics originating from fisheries is difficult to estimate, and the literature reports very little on the subject. However, a recent report from the University of Plymouth states that the total number of microplastic fragments in the oceans originating from the use of fishing gear in the United Kingdom can range from 326 million to 17 billion pieces annually [20]. Another study from the University of Alicante shows that the concentration of microplastics in marine sediments on the coast of Spain is higher close to the three coastal fish farms investigated [21]. There are also other sources of plastic pollution caused by fishing that we do not consider, such as lost and abandoned gear that remains in the ocean indefinitely. Lost and abandoned gear causes severe problems, such as ghost fishing [22] and entanglement [2]. A recent report claims that lost and abandoned fishing gear contributes to more than 45000 tons of plastic pollution annually [23].

The objective of this chapter is to determine the number of microplastics in the sea originating from fishing gear globally. Due to the complexity, it must be considered a first approach, aiming for better and more accurate calculations in the future. We omit small-scale fishing and consider commercial fishing only, that is, fishers having a quota and regularly delivering catch registered in the catch statistics.

2. Methodology

2.1 Causes of wear on fishing gear

The leading causes of fishing gear wear are abrasion with the sea bottom and hauling equipment. Specifically, ropes and nets dragged along the seafloor create heavy wear on trawls and Danish Seine ropes, and for these gear types, it is common to change the parts that are in contact with the sea bottom after 1–2 years and in some cases even more often. The extent of wear depends on the seafloor condition, meaning that a rocky bottom creates much more wear than a sandy bottom.

Another significant cause of abrasion is the onboard hauling equipment. Hydraulic net haulers, as shown in **Figure 4**, or net drums are the two most common haulers in use, and for both, the heavy stress caused by the ropes pressing toward the equipment causes abrasion that gradually tears down the ropes. For the net hauler, the rope is squeezed between two plates, creating even more stress on the rope.

In addition to the gradual abrasion caused by contact, plastic ropes also get degraded by other causes. The most common is UV radiation, leading to the fragmentation of the plastic fibers. Fishing gear is constantly exposed to UV radiation, and proper storing of the gear is essential to prevent degradation. In addition, fragmented plastic ropes have accelerated wear when in contact with the hauler or the seafloor. According to a study from the United Kingdom, plastic ropes lying in the sea at 10 m depth lose an average of 0.39% (PP), 1.02% (PA), and 0.45% (PE) per month caused by abrasion due to UV radiation [24].

Also, gear dragged in the water wears due to the friction force between water and gear. This effect is much less than bottom contact but still significant, combined with the fragmentation effect caused by UV radiation. However, dragging occurs mainly during hauling, and the hauling equipment is considered a much more significant source of depletion.

2.2 Methodological approach

The research and reports on the wear and tear of fishing gear are scarce. Thus, we need to establish a methodology we can use to approach the solution. Due to the complexity, it requires an enormous effort to get exact numbers of microplastics generated from fishing gear. Instead, our goal is to get a rough estimate of the number of microplastics in the sea to get a feeling of how severe the problem is and determine which fishing gear causes the most pollution.

For each fishing gear considered, we interviewed five fishers to get basic information on their use of the gear. This information includes the average number of gear in use, the length of the ropes, the average lifetime, and their estimate of the wear and tear when the gear is replaced. In addition, we checked some of these figures against the sales figures from leading gear manufacturers. Furthermore, the Norwegian statistics for fisheries [25] provide the number of vessels for each fishing gear and their total catch.

Additionally, we collected samples from dispatched gear, mainly seine ropes and longlines, and measured the diameter and weight of the ropes to calculate the depletion. These calculations were then compared with the information from the fishers, further providing a better estimate of the wear and tear. The result gives the average percentage of wear for each type of fishing gear. We then calculate the total wear and tear by finding the total number of gear used.

Expanding the scope to include all fishing gear worldwide is indeed a challenging exercise. Unfortunately, an overview of all fishing vessels and their gear is not readily available, nor are the conditions for their fishing. Therefore, we must settle for rough estimations. We have statistical data for the global catch produced by FAO since 1950 [26], and our first approach is to use this data and assume the same conditions apply to other nations than Norway. In this way, we assume the amount of microplastics generated per ton of fish is equal for every nation. Using the statistical data from FAO [26], we find that Norway accounts for about 3.0% of all catches worldwide, which is a starting point for our calculations.

3. Danish seine

In a recent report, we have described the plastic pollution caused by the Danish seine fishery [27]. Therefore, we do not go into detail but briefly describe this fishing gear and the main findings. For details, we refer to [27].

3.1 Usage and causes of wear

The Danish seine comprises a conical net with rope arms at each side. The ropes usually have a steel wire core since they must withstand heavy forces when dragged along the seabed. Several variants of this fishing method exist, including the original Danish seine method, called Anchor seining. Other methods are fly-dragging, also called Scottish seine, and tow-dragging, also called the Japanese method. For all of these variants, the rope arms are dragged along the seabed while the net wraps around the catch. The rope arms may be several kilometers in length. The sea bottom contact tears heavily on the ropes, and usually, the fishers replace them after 18 months. This replacement rate may be specific for Norway and possibly differ for other countries' fisheries.

3.2 Calculated wear

By comparing the weight of new and used seine ropes, we estimate the loss due wear and tear. Furthermore, based on interviews with the fishers, we get statistics on the ropes used. We calculate the annual wear on the Danish seine fishery in Norway to be 77–97 t plastic. These figures depend on several factors, such as the average lifetime, sea bottom conditions, stretching of the ropes, and the average rope arm length and number of seine in use. In other words, there are many possible sources of error, but we believe we are close to the actual value. Then, by finding the number of vessels in use for other countries and assuming the same wear is also valid, we estimate the annual worldwide plastic pollution from seine fishery to be about 311 t [27].

4. Bottom trawl

4.1 Usage and causes of microplastics

Bottom-trawl fishing, i.e., beam trawls, otter trawls, and dredges, is used worldwide and provides about a quarter of the marine catch [28]. A rich body of literature assesses the impacts of sea bottom trawling on benthic invertebrate disturbances and seabed alterations [29, 30]. However, research on the wear and tear of sea bottom trawl remains relatively understudied.

In Norway, the rockhopper ground gear has been commonly used in the sea bottom trawl fisheries for the last 30 years. In contrast to the previously used bobbin ground gear, the rockhopper has shown improved catch efficiency for Haddock (*Melanogrammus aeglefinus*) and Atlantic Cod (*Gadus morhua*) due to its increased contact with the seabed [31].

The trawl comprises trawl doors, bridles, sweeps, and ground gear (rockhopper). The fish is caught by dragging the trawl net along the seabed. The water is released through the net mesh, and the fish remains in the trawl bag. The ground gear consists of steel and rubber, while the wipers and doors consist of steel. Dolly ropes or other protective ropes used to protect the net consist of bundles of plastic (PE) threads. The ground gear discs are made from old dump truck tires and threaded onto heavy wires or chains. The most common are discs of 21 inches in diameter, but on the rough seafloor or when fishing for halibut, the discs can be 24 inches. The material composition of a standard dump truck tire is approximately one-third natural rubber, while the rest is a mixture of synthetic rubber and filler.

The lifespan of a sea bottom trawl depends on several factors, such as seabed composition, trawl type, traction speed, and local hydrodynamic forces (i.e., current). However, the seabed condition, like the roughness, is the decisive factor in how quickly the trawl components wear. Areas with fungus, followed by stone or rocky bottoms, are abrasive, while clay is the most gentle seafloor type. It is difficult to estimate the wear and tear on the trawl net itself. The line is rarely so worn that it peels off. Before this occurs, sections are exchanged or the line repaired. Fishers replace the protecting net regularly, but exact replacement rates are challenging to obtain from fishers and manufacturers. However, it is possible to estimate wear on the ground gear, which the fishers replace when worn out.

4.2 Calculated wear

To calculate the annual mass loss from the rockhopper gear, L_{rh}, we use the following equation:

$$L_{rh} = N_{disc} W_{disc} N_{tw} N_v P_w \frac{12}{LT} \tag{1}$$

Here, N_{disc} is the number of discs at the rockhopper, W_{disc} is the weight of each disc, N_{tw} is the number of trawls in use, N_v is the number of vessels, P_w is the average percentage wear when replaced, and LT is the average lifetime in months (**Figure 1**).

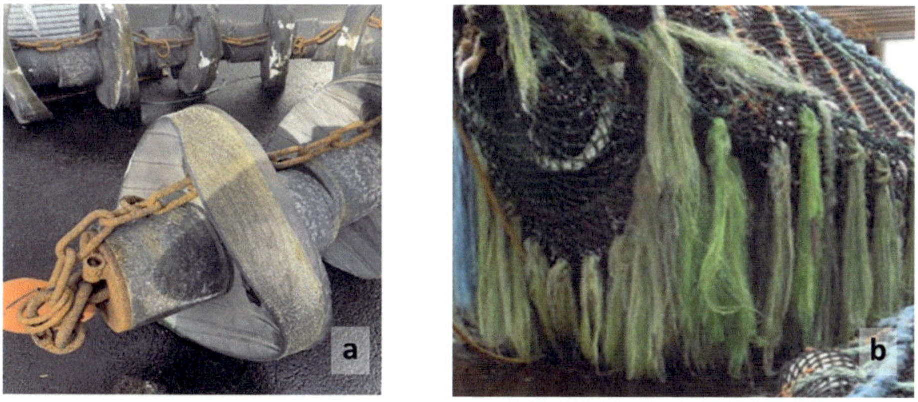

Figure 1.
Worn-out rockhopper gear (a) and dolly rope used to protect the trawl (b). Photo: SINTEF.

In conversations with Norwegian shipping companies and gear manufacturers, Norwegian vessels mainly use 21-inch discs weighing between 20 and 23 kg, depending on the manufacturer. Thus, in the calculations, we used a weight per disc W_{disc} = 21 kg. Gear has five sections, two starboards, two ports, and one center gear. Each section consists of 21 discs, giving a total of N_{disc} = 105.

A rockhopper gear is touching the seafloor in all its length and thus causing massive depletion of the gear, see **Figure 1a**. The manufacturers state that the gear is replaced every 6–10 months, at which they are approximately 20% worn. However, the fishers in our study claim the gear last longer, approximately 12–18 months, but with more significant wear, usually 30–40%. In our calculations, we make a sober estimate of the lifetime to 10 months for a standard gear with a percentage weight loss of 20%. The wear is vessel-specific, so these numbers must be considered average for the fleet.

We consider only vessels that delivered more than 100 t catch during 2019, in which case the number of vessels is 63. Half of them use twin trawl. **Table 1** summarizes the calculations on wear and tear from rockhopper gear, which is close to 50 t annually.

The protecting ropes, like dolly ropes, are also heavily exposed to wear and tear. A Dutch research consortium, DollyRopeFree, has reported weight loss of 10–25% after 2 weeks of use [32]. Unfortunately, they do not mention from which type of trawl the samples were taken. We have not been able to calculate the annual plastic fragmentation from dolly ropes due to a lack of information. However, there is no doubt that dolly ropes and other protective ropes contribute to plastic debris in the ocean on a large scale. A worn dolly rope is shown in **Figure 1b**.

	N_{discs}	W_{discs}	N_{tw}	N_v	P_w	LT	L_{rh}
Single trawl	105	21 kg	1	32	20%	10 months	16.9 t
Twin trawl	105	21 kg	2	31	20%	10 months	32.8 t
Total							49.7 t

Table 1.
Calculated annual mass loss from rockhopper gear.

4.3 Trawling worldwide

Unfortunately, there are no registrations on trawlers worldwide, making it difficult to estimate the global share of microplastics from this gear. Also, the catch statistics are difficult to use since they are sorted by species, not gear type. Therefore, acknowledging that the Norwegian catch share is about 3% is the only way, we can now estimate the global loss. Hence, using the Norwegian numbers for trawl and dividing by the Norwegian share, we end up with a total global microplastic loss due to trawls of 1 656 t.

5. Gillnets

5.1 Usage and causes of microplastics

Gillnets are the most important commercial fishing gear for the Norwegian coastal fleet, where cod (*Gadus morhua*) and saithe (*Pollachius virens*) are the most important target species. In 2019, 74 864 t cod and 32 032 t saithe were caught using this gear.

A gillnet consists of the float line, the lead line, and the netting. The float and lead lines usually consist of polypropylene and polyethene, but the lead line also has lead inside the core to make it sink. In addition, a drop line connects to a buoy at the sea surface. The drop line consists of polypropylene and polyethene at the lower part and polyester at the upper part. The netting consists of single or multistrand monofilament nylon.

The lifetime of a gillnet is highly variable and based on geographical location, vessel size, and hauling frequency, which in turn depends on the type of fishing. For example, some vessels fish all year round and replace a thousand nets yearly. Others, who only operate during the cod season, may replace the net each 4–5 years, or even after10 years. Also, as for the trawl, there are significant variations in the lifetime depending on the seabed quality.

Drop and float lines are rarely in contact with the seafloor, and abrasion mainly occurs during hauling and setting. Although it is challenging to calculate wear and tear solely from the hauling equipment, gillnet fishing at great depth causes more significant stretches and squeezes during hauling than fishing in relatively shallow waters. Blue halibut is one of the species caught at such great depth. Lead lines, on the other hand, are more prominent to wear and tear due to their contact with the sea bottom. **Figure 2** shows an example of a worn-out lead line (a) and a typical gillnet hauler (b). The ropes are squeezed between two plates at the hauler, creating

Figure 2.
Worn-out lead line (a) and a typical gillnet hauler (b). Photo: SINTEF.

significant abrasion. Another common type of hauler is the drum winch, which also creates abrasion of the ropes but is much more gentle.

5.2 Calculated wear and tear

To estimate the wear and tear on gillnets, we need knowledge of several parameters, such as the average number of gillnets per vessel, the thickness of the ropes, the average lifetime, and the average percentage mass loss. We got these numbers from our interviews with fishers, confirmed by information from gear suppliers.

On average, a fisher has 200 gillnets consisting of netting, lead, and float lines. They are tied in strings of different lengths, where the length of each component is 27.5 m. The total length (L_G) of all 200 gillnets is 5500 m. In addition, a fisher has 2000 m of drop lines on average. For the netting, the weight is 2.2 kg for one net, i.e., the complete 27.5-meter length, giving a weight per meter (W_m) of 80 g. For the drop line, 14 mm is the typical diameter, weighing 88 g per meter. For the float line, a 15 mm rope is used with a weight of 100 g per meter, and for the lead line, a 12 mm diameter is used with a weight of 73 g per meter, excluding the lead.

Next, we have to estimate the annual loss percentage. The fishers estimate a percentage loss, P_L, of about 8% when the ropes are worn out, and for the average lifetime (LT), they estimate 15 years for lead lines, 20 years for float lines, and 25 years for drop lines. The numbers are based on an average of 40–60 trips per year. The netting is replaced every 4 years on average. The netting often gets stuck in rocks at the sea bottom, creating holes and small pieces that are torn apart. Due to this, it must be replaced more often. We estimate a loss percentage of 3%, giving an annual loss percentage of 0.75%. To calculate the annual loss (L) for each component, we then use the following formulae:

$$L = N_V L_G W_m P_L / LT \qquad (2)$$

Table 2 shows the calculated loss per component and vessel and the total loss per fleet. According to the Norwegian statistics for fisheries [25], the number of vessels using gillnets is N_V = 1472.

5.3 Gillnets worldwide

Gillnets are perhaps the most used fishing gear worldwide. However, in many areas, there are differences in use from the Norwegian method described above.

	L_g (m)	W_m (kg)	W_G (kg)	P_L	LT (years)	L_V (kg)	L (kg)
Lead lines	5 500	0.073	401.5	15%	15	4.02	5 910
Float lines	5 500	0.1	550	15%	20	4.13	6 072
Drop lines	2 000	0.088	176	15%	25	1.06	1 554
Netting	5 500	0.08	440	3%	4	3.3	4 857
Total							18 394

Table 2.
Calculated annual microplastics loss from gillnets per vessel and total per fleet.

Smaller boats are typical, with other types of equipment and ropes with other dimensions. Therefore, estimating the global amount of microplastics based on Norwegian numbers is very challenging. Our approach is to use the Norwegian numbers and divide them by the estimated Norwegian share of the total catch. Unfortunately, the statistics are grouped on species, not gear type, meaning we cannot know the Norwegian catch share for a specific gear type. To overcome this problem, we assume the catch share for gillnets is the same as the total catch share for all gears in use. This approach is a gross simplification, which we have to keep in mind.

According to FIGIS [26], Norway accounts for 3.0% of the total catch of marine and diadromous species. Thus, the global amount of microplastics generated from gillnets estimates to 613.1 t.

6. Pots

6.1 Usage and causes of microplastics

Crab pots are tied in strings, but the number of pots in each string varies on the type of fishery. They come in many different shapes, but the basic idea is to trap the crabs inside the pots. King crab pots are collapsible to ease storage, and Snow crab pots are conical so that they can be stacked.

In Norway, we have identified 426 vessels catching crabs based on the criteria that they have landed more than 400 kg per year. On average, we assume 15 strings with 15 pots per string, based on the interviews with fishers. The pots usually stand at 30–40 m depth with 20 m spacing. In addition, we have 772 vessels fishing for King crab in the northern part of Norway. For King crab, the depth is about 200 m, which causes more stress and more wear on the ropes during hauling. The typical rope diameter is 10 mm. There are different rope qualities, mostly polypropylene, polyethene, or nylon.

Fishing for snow crab takes place in the Barents Sea (Norwegian and Russian fishers) and the Northwest Atlantic and North Pacific, usually at depths of 220–300 m. A large number of pots in the string is typical for snow crab fishery, usually 200 on average. The ropes typically have a diameter of 22–24 mm due to the heavy stress they are exposed to. Thus, the rope thickness and number of pots are unique for this type of fishery. The pots have a conical shape, as seen in **Figure 3a**. A vessel fishing for snow crab may have 35 strings and thus a total of 7500 pots.

The hauling equipment is the most significant cause of wear and tear on the drop line and the connecting ropes between the pots. Ropes are hauled quickly from a depth of 220–300 m, causing significant abrasion. The winch also contributes to wear and tear on the ropes, pulling the ropes backward and through the boat to the bins behind.

Also, the plastic coating around the steel cracks during use, but there are significant differences in the quality and how much it cracks. Some of the cheaper pots are of low quality and tend to rust. When the iron rusts, the plastic coating explodes, as seen in **Figure 3b**, releasing large plastic flakes into the sea. We do not include the plastic originating from this coating in our calculations.

6.2 Calculated wear

To estimate the wear on the ropes, we distinguish between Snow crabs and other crabs and lobsters. Then we find the number of vessels involved in both categories and estimate the length of the ropes based on the average number of strings and pots

Figure 3.
Snow crab pot (a), with cracked coating (b). Photo: SINTEF.

in each string. Finally, we estimate the average wear percentage based on fishers' and manufacturers' information.

For crabs and lobsters, the total number of vessels involved in Norway is 1198, half of them catching King crab. The average number of pots at each vessel depends on the type of crab, but we use 100 pots divided into 10 strings with 10 pots each. The distance between each pot is 20 m, and the drop line length is 50 m on average.

For snow crabs, there are only nine active vessels. After dialogue with the fishermen, we estimate an average of 7000 pots, spread over 35 strings with 200 pots in each string. The typical distance between each pot is 30 meters. The drop line at each end of the string is usually three coils of 110 meters.

To calculate the total length of ropes (L_R) for the whole fleet, we use the following formula:

$$L_R = N_V N_s \left[L_d + \left(N_p - 1 \right) L_i \right] \tag{3}$$

Here, N_V is the number of vessels, N_s is the average number of strings, N_p is the number of pots in a string, L_d is the dropline length, and L_i is the rope length between the pots in the string. The above information is summarized in **Table 3**, where the length calculations are based on Eq. (3).

Finally, we calculate the annual mass loss (L) due to wear and tear from Eq. (4):

$$L = W_m L_R P_L \tag{4}$$

Here, W_m is the rope weight per meter, and P_L is the annual percentage loss. Finally, W_T in **Table 4** is the total rope weight. In particular, the annual loss percentage is challenging to estimate. However, the results from [24] can be used as a starting point. The ropes are mainly PP/PE, and [24] suggests a monthly mass loss of 0.4% for

Category	N_V	N_P	N_S	L_i	L_d	L_R (m)
Snow crab	9	200	35	30	660	2 088 450
Other crabs	1198	10	10	20	50	2 755 400
Lobster	50	5	10	20	50	65 000

Table 3.
Calculation of the total rope lengths in meters for crab fishery in Norway.

Category	W_m (kg)	L_R (m)	W_T (kg)	P_L	L (kg)
Snow crabs	0.259	2 088 450	540 909	2.5%	13 523
Other crabs	0.045	2 755 400	123 993	1.5%	1 860
Lobster	0.045	65 000	2 925	1.5%	44
Total					15 427

Table 4.
Calculation of the annual amount of microplastics due to crab fishing in Norway.

such ropes just by lying in the sea. Additionally, the haulers contribute significantly to this loss. Since the ropes are not constantly in the sea, and their lifetime often is 10 years or more, we estimate an annual loss percentage of 1.5% for crab and lobster pots. The loss is set to 2.5% annually for snow crab ropes since the stress is much higher. The results are then summarized in **Table 4**.

6.3 Pots worldwide

Crabs and lobsters are caught all over the world. Asia is the most significant area, with China and Indonesia as the leading nations. For lobster, Canada is the primary nation.

In Norway, King crab and Snow crab (Queen crab) represent a significant source of income for the Norwegian seafood industry and are commercially more valuable than other crab types. Worldwide, Canada is the leading nation in fishing for Snow crab, but the USA, Japan, Russia, and Greenland are also active nations. However, there are significant differences in the use of snow crab pots compared with other pots, as previously explained; hence, we consider them separately. **Table 5** shows the crab catch in 2019 divided by type and area. Worldwide, snow crabs account for only 13% of the total crab and lobster catch.

The total catch of Snow- or Queen crabs worldwide is 116 748 t. Out of this, Norway accounts for 3.5%. The total catch for crabs is 1 461 581, and the majority is from Asia. Norway accounts only for 0.5% of the total crab catch. Finally, the total lobster catch is 320 057 t, where Norway accounts for 0.1% of the catch.

To estimate the total global generated microplastics from crab fishery, we use data from **Table 5** and divide by the Norwegian share. The results are shown in **Table 6**.

In Norway, fishing for Snow crabs is the main contributor to microplastics, even with a small number of vessels involved. However, other crabs account for the more significant part of the wear and tear worldwide. The global amount of microplastics originating from the crab fishery is 802.2 t.

	Total	Africa	Asia	Oceania	America	Europe	Norway
Snow (Queen) crab, catch in tons	116 748		13 200		89 678	13 870	4 049
Other crabs, catch in tons	1 461 581	42 380	1 085 526	1 762	180 072	151 840	7 078
Lobster, catch in tons	320 057	16 966	15 907	13 845	208 041	65 298	433

Table 5.
Crab catch in 2019 divided by type and area — derived from [26].

Category	Microplastic loss in Norway (kg)	Norwegian share	Annual worldwide microplastic loss (kg)
Snow crab	13 523	3.5%	386 363
Other crabs	1 860	0.5%	371 979
Lobster	44	0.1%	43 875
Total	15 427		802 217

Table 6.
Annual worldwide microplastics generated from crab fishery.

7. Longlining

In longline fishing, we distinguish between three different modes of operation: bottom longline, surface longline, and autoline. In Norway, autoliners in the high seas account for 70% of the total catch. This fleet consists of 20 vessels, and the total annual catch in 2019 from this segment was 57 428 t. Target species are demersal fish, such as cod and saithe.

A longline consists of a long rope (the mainline) attached to branch lines with hooks. The mainline is a multifilament line (polyester and polyamide) consisting of three or four cord sections twisted together to form a long rope. The length can be up to 180 kilometers. Depending on the fishery, the mainline generally ranges from 4 to 11 mm in diameter. The supply is made of polyester, but the hooks, swivels, and stoppers are steel; see **Figure 4a**.

7.1 Usage and causes of microplastics

Most of the wear and tear occur during hauling. Hauling is performed by a powered line hauler, where the rail roller guides the longline over the rail of the ship before the de-hooker and hook cleaner remove the fish and unused bait from the hooks. Furthermore, the twist remover takes out the twist in the line before the hook separator guides the hook into the storage rack, where they are held in magazines.

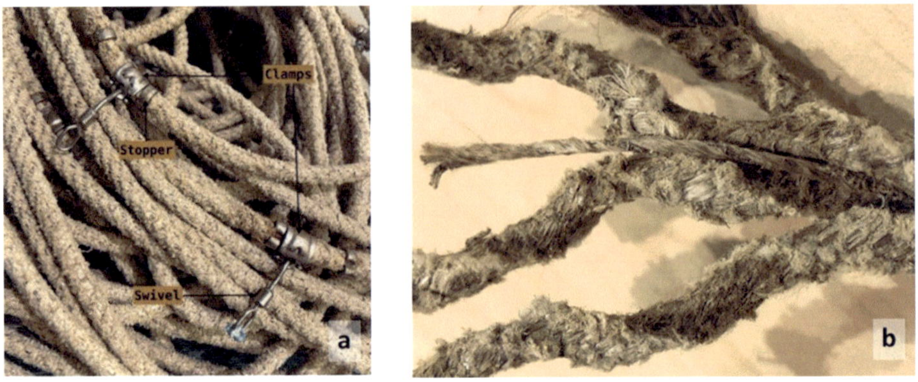

Figure 4.
Autoline with stoppers, swivels, and clamps (a) and a worn-out autoline with the cords split apart (b). Photo: SINTEF.

On a traditional autoline vessel, the hauling equipment is located on the vessel's starboard side. However, there has been a shift toward hauling through moonpools in the center of the vessels. As the moonpool is placed with the lowest magnitude, it may reduce wear on the line. However, this is only an assumption as no such research exists in the area.

7.2 Calculated wear

To calculate the annual mass loss due to wear and tear from longlines (L), we use the following formulae:

$$L = N_V N_l L_l W_m P_L \,/\, LT \tag{5}$$

N_V is the number of vessels in the fleet, which we get from the Norwegian statistics for fisheries [25]. N_l is the average number of lines for each vessel, L_l is the average length of each line, W_m is the average weight per meter, P_L is the average percentage loss when the line is worn-out, and LT is the average lifetime. All of these parameters were acquired through our interviews with fishers.

Additionally, we got one worn-out line, **Figure 4b**, which we measured to establish the mass loss. This line has been in operation for 1.5 years and has been hauled approximately 300 times. A new line is 180 m long and weighs 16 kg. Our worn-out line weighs 14.9 kg, thus reduced by 1.9 kg. This gives a percentage loss of 11.9%, which coincides with what our fishers told us (12%). Measuring only one line does not give a statistically sound result, but it underpins the statements from the fishers.

Table 7 shows the calculated loss for longlines and the parameters used for the calculations Eq. (5). We split the lines into four categories as each category's parameters differ. The total loss from longlines sums to 37.2 t annually.

7.3 Longlining worldwide

As for gillnets, we cannot provide well-founded numbers on the usage of longliners worldwide. Typically, longlines are used to catch tuna in Japan, Taiwan, Korea, Cuba, and Oceania. Other nations use longlines to catch halibut in the northern

	N_V	N_l	L_l (m)	W_m (kg)	P_L	LT (years)	L (kg)
Autoline, high seas	20	500	180	0.078	12%	1.5	11 232
Autoline, coastal	53	80	180	0.100	12%	1.5	6 105
Bottom longline	415	120	540	0.035	8%	4	18 824
Surface longline	103	120	540	0.013	5%	4	1 042
Total							37 205

Table 7.
Calculated annual mass loss for different types of longlines.

Pacific. Therefore, our best estimate is to use the method for gillnets, assuming the Norwegian share of worldwide longline use is about 3%. In that case, the total microplastics from longlining worldwide is 1 240 t.

8. Discussion

8.1 On the methodology

The methodology is based on interviews with fishers to acquire essential parameters we need to calculate loss from gear. The annual loss percentage is a crucial parameter we have tried to check using different methods, such as measuring worn-out ropes and using numbers from the literature [24]. However, the many uncertainties are a weakness of the study. To get better estimates, we believe performing comprehensive studies on different types of worn-out gear is the way forward.

For calculating the global wear and tear, we need access to the number of gears used globally for each gear type. This may be available for some countries but is very hard to find. In addition, statistical parameters, such as rope lengths, diameters, and average lifetime, are necessary to perform reliable calculations for different areas. We acknowledge that our methodology for finding the global wear and tear on fishing gear has many weaknesses, but on the other hand, it is the best we can do for now. Improving the methodology is an important area for future work.

8.2 The results

Table 8 summarizes the worldwide microplastics generated from the wear and tear of the different fishing gear we have considered, together with the Norwegian numbers. We have used the Norwegian fishing fleet as the basis for our calculations. Although the Norwegian numbers are uncertain, we believe they, in general, are close to reality, considering that the uncertainties may equalize each other.

The UK study [20] concludes with 326 million to 17 billion fragments of microplastics from the UK fishery. It would be great to compare this to our calculations, but unfortunately, the size and weight of a fragment are undefinable. However, their measurements show the mass loss from haulers ranges from 12 μg to 1050 μg per meter hauled for new and 10-year-old ropes, respectively [20], indicating that older

Fishing gear	Calculated annual microplastics from the Norwegian fishery (tons)	Calculated annual microplastic from the global fishery(tons)
Danish seine	87	311
Bottom trawl	49.7	1 656
Gillnet	18.4	613
Longline	37.2	1 240
Crab pots	15.4	802
Total	207.7	4 622

Table 8.
Summary of the Norwegian and global microplastics generated from wear and tear on fishing gear.

ropes wear more quickly than new ones. Therefore, we calculated the mass loss per meter rope hauled based on our calculated mass loss in Norway for some gear components worn primarily due to the hauler. The results are shown in **Table 9**, indicating that for surface longlines our results are comparable to [20], whereas, for gillnets and crab pot ropes, we are possibly underestimating the wear. Another explanation is that the haulers used for crab pots and gillnets are of the drum types that are more gentle to the ropes than the ones tested by [20].

For Danish seine, we have a reasonable understanding of the number of vessels involved worldwide; hence, the total numbers are well founded. Also, the wear and tear from the Norwegian fleet are based on interviews with fishers backed up by measurements on worn-out ropes. Thus, the numbers presented for the Danish seine are probably the ones best founded.

For trawls, we only consider sea bottom trawls as these are the ones contributing the most to the pollution. We do not consider Shrimp trawls as the demanded information is more challenging to get, and these trawls do not have plastic parts in contact with the seafloor. Yet, dolly ropes and other protective ropes are dragged along the seafloor, contributing considerably to the total plastic pollution. Hence, we believe that the numbers we present for trawl pollution are highly underestimated. The microplastic pollution originating from dolly ropes and other protective ropes is undoubtedly a topic that needs further investigation. Also, we do not have numbers for the worldwide use of trawlers; this is also a field in which more research is needed.

For gillnets, we use average numbers for the number of nets per fisher. These are based on interviews with fishers but are still questionable as the fishing fleet is large, and there may be significant differences between fishermen. Further differences are introduced when we consider gillnet fishing worldwide. Fishers in many countries probably use other types of ropes than Norway, and the average number of nets is

	Annual mass loss (kg)	Total length (m)	Average hauls per year	Mass loss per meter hauling (μg)
Gillnet float line	4.13	5 500	50	15.0
Crab pot ropes	1 860	2 755 400	20	33.8
Surface longline	1 042	64 800	70	229.7

Table 9.
Calculated mass loss per meter hauling for some gear components.

tough to estimate. Hence, the amount of microplastics from global gillnet fishing is a very rough estimate, and better statistics on using gillnets globally are needed.

For pots, we split between snow crab and other types of pots due to the different operating modes. As for the other gears considered, there are many uncertainties in the parameters used for the calculations, with the percentage loss being the most dubious. However, the numbers for Norway are considered well founded. For the global loss, there are probably significant differences in how the fishery is performed, and the average amount of pots and ropes used varies from country to country. Therefore, our estimates for the global loss are highly uncertain, and more research is needed to establish more accurate numbers.

Also, the main uncertainty for longlining is establishing a correct number for the annual percentage loss. Our fishers have estimated this to be 8%, which may be correct but difficult to verify. We don't have information on the global number of vessels and the parameters for each type of line. Hence, our numbers for the global loss are based on loose assumptions. More research must be performed on the global use of longlines to get more accurate results.

9. Conclusions

Our results show that the number of microplastics originating from fisheries worldwide is 4 622 t, which is probably a conservative estimate. We omitted some gears, like ordinary seine, and we only considered the rockhopper gear for the trawl. Hence, dolly ropes and trawl mats are not included, and we believe they contribute significantly to the total amount. The number of microplastics from lost and abandoned fishing gear is estimated to be 45000 tons [23]. This number is almost 10 times higher than our calculation but can make sense since lost and abandoned gear are complete, not only fragments. If correct, this suggests that the effort should be put into avoiding such lost and abandoned gear. However, we believe that finding ways to reduce microplastic wear and tear from commercial fishing activities is an important task too.

Despite many uncertainties, our calculations and results can provide helpful information and are essential to highlight the topic of microplastics originating from ordinary fishing activities. Then, we believe future research will lead to more accurate numbers. Specifically, more research on the global use of longlines, gillnets, and crab pots is needed, and a better understanding of dolly ropes and trawl mats is essential since these components contribute significantly to the total pollution.

Acknowledgements

Jørgen Vollstad at SINTEF Ocean has contributed to the work by sharing knowledge of the fisheries in Norway and helped to find the underlying numbers for the calculations. Bård Johan Hanssen at SINTEF Nord has helped to find statistics for the number of fishers and catches in Norway.

The Norwegian Directorate of Fisheries has partly supported this work.

Author details

Tore Syversen* and Grethe Lilleng
SINTEF Nord, Tromsø, Norway

*Address all correspondence to: tore.syversen@sintef.no

References

[1] Andrady AL. Microplastics in the marine environment. Marine Pollution Bulletin. 2011;**62**(8):1596-1605

[2] Derraik J. The pollution of the marine environment by plastic debris: A review. Marine Pollution Bulletin. 2002;**44**:842-852

[3] Cole M, Lindeque P, Halsband C, Galloway TS. Microplastics as contaminants in the marine environment: A review. Marine Pollution Bulletin. 2011;**62**(12):2588-2597

[4] Ballerini T, Pen JR, Andrady A, Cole M, Galgani F, Kedzierski M, et al. Plastic pollution in the ocean: What we know and what we don't know about. 2018

[5] Thevenon F, Carroll C, Sousa J, editors. Plastic Debris in the Ocean: The Characterization of Marine Plastics and Their Environmental Impacts, Situation Analysis Report. Gland, Switzerland: IUCN; 2014

[6] Hammer J, Kraak M, Parsons J. Plastics in the marine environment: The dark side of a modern gift. Reviews of Environmental Contamination and Toxicology. 2012;**220**:1-44

[7] Pruter AT. Sources, quantities and distribution of persistent plastics in the marine environment. Marine Pollution Bulletin. 1987;**18**:305-310

[8] Thushari GGN, Senevirathna JDM. Plastic pollution in the marine environment. Heliyon. 2020;**6**(8): e04709

[9] Abbasi S, Soltani N, Keshavarzi B, Moore F, Turner A, Hassanaghaei M. Microplastics in different tissues of fish and prawn from the Musa Estuary. Persian Gulf; 2018

[10] Bessa F, Barría P, Neto JM, Frias JPGL, Otero V, Sobral P, et al. Occurrence of microplastics in commercial fish from a natural estuarine environment. Marine Pollution Bulletin. 2018;**128**:575-584

[11] Neves D, Sobral P, Ferreira JL, Pereira T. Ingestion of microplastics by commercial fish off the Portuguese coast. Marine Pollution Bulletin. 2015;**101**(1):119-126

[12] Bellas J, Martínez-Armental J, Martínez-Cámara A, Besada V, Martínez-Gómez C. Ingestion of microplastics by demersal fish from the Spanish Atlantic and Mediterranean coasts. Marine Pollution Bulletin. 2016;**109**(1):55-60

[13] Devriese LI, van der Meulen MD, Maes T, Bekaert K, Paul-Pont I, Frère L, et al. Microplastic contamination in brown shrimp (Crangon crangon, Linnaeus 1758) from coastal waters of the Southern North Sea and Channel area. Marine Pollution Bulletin. 2015;**98**(1-2):179-187

[14] Lusher AL, McHugh M, Thompson RC. Occurrence of microplastics in the gastrointestinal tract of pelagic and demersal fish from the English Channel. Marine Pollution Bulletin. 2013;**67**(1-2):94-99

[15] von Moos N, Burkhardt-Holm P, Köhler A. Uptake and effects of microplastics on cells and tissue of the blue mussel Mytilus edulis L. after an experimental exposure. Environmental Science & Technology. 2012;**46**(20): 11327-11335

[16] Jabeen K, Su L, Li J, Yang D, Tong C, Mu J, et al. Microplastics and mesoplastics in fish from coastal and fresh waters of China. Environmental Pollution. 2017;**221**:141-149

[17] Possatto FE, Barletta M, Costa MF, et al. Plastic debris ingestion by marine catfish: An unexpected fisheries impact. Marine Pollution Bulletin. 2011;**62**(5):1098-1102

[18] Hossain MS, Rahman MS, Uddin MN, Sharifuzzaman SM, Chowdhury SR, Sarker S, et al. Microplastic contamination in Penaeid shrimp from the Northern Bay of Bengal. Chemosphere. 2020;**238**:124688

[19] Hall NM, Berry KLE, Rintoul L, Hoogenboom MO. Microplastic ingestion by scleractinian corals. Marine Biology. 2015;**162**(3):725-732

[20] Napper IE, Wright LS, Barrett AC, Parker-Jurd FNF, Thompson RC. Potential microplastic release from the maritime industry: Abrasion of rope. Science of The Total Environment. 2022;**804**:150155

[21] Krüger L, Casado-Coy N, Valle C, Ramos M, Sánchez-Jerez P, Gago J, et al. Plastic debris accumulation in the seabed derived from coastal fish farming. Environmental Pollution. 2020;**257**:113336

[22] Macfadyen G, Huntington T, Cappell R. Abandoned, Lost or Otherwise Discarded Fishing Gear. Rome: United Nations Environment Programme: Food and Agriculture Organization of the United Nations; 2009. p. 115

[23] Kuczenski B, Vargas Poulsen C, Gilman EL, Musyl M, Geyer R, Wilson J. Plastic gear loss estimates from remote observation of industrial fishing activity. Fish and Fisheries. 2022;**23**(1):22-33

[24] Welden NA, Cowie PR. Degradation of common polymer ropes in a sublittoral marine environment. Marine Pollution Bulletin. 2017;**118**(1):248-253

[25] Norwegian Directorate of Fisheries. Statistics for fisheries. 2022. Available from: https://www.fiskeridir.no/English/Fisheries/Statistics

[26] FAO. FIGIS - Global capture production 1950-2019. 2022. Available from: https://www.fao.org/figis/servlet/TabSelector

[27] Syversen T, Lilleng G, Vollstad J, Hanssen BJ, Sønvisen SA. Oceanic plastic pollution caused by Danish seine fishing in Norway. Marine Pollution Bulletin. 2022;**179**:113711

[28] Amoroso RO, Pitcher CR, Rijnsdorp AD, McConnaughey RA, Parma AM, Suuronen P, et al. Bottom trawl fishing footprints on the world's continental shelves. Proceedings of the National Academy of Sciences. 2018;**115**(43):E10275-E10282

[29] Hiddink J, Jennings S, Sciberras M, Bolam S, Cambie G, Mcconnaughey R, et al. Assessing bottom-trawling impacts based on the longevity of benthic invertebrates. Journal of Applied Ecology. 2018;**56**

[30] Jones J. Environmental impact of trawling on the seabed: A review. New Zealand Journal of Marine and Freshwater Research. 1992;**26**:59-67

[31] Engås A, Godø O. Escape of fish under the fishing line of a Norwegian sampling trawl and its influence on survey results. Journal du Conseil International pour l'Exploration de la Mer. 1989;**45**:269-276

[32] Wouter J. Cleaning the Nordic seas from dolly rope. DollyRopeFree. 2018. Available from: http://www.dollyropefree.com/blog/23/cleaning+the+nordic+seas+from+dolly+rope

Chapter 9

Microplastic Contaminants in the Sediment of the East Coast of Saudi Arabia

Hamza Jawad Al-Shaikh Ali, Assad Ahmed Al-Thukair, Abdurahiman Kambrath Pulikkoden and Basheer Chanbasha

Abstract

Microplastic contamination in the sediment of the east coast of Saudi Arabia was not addressed by any study. The objective of this study is to obtain the first measurement of microplastic abundance at four different beaches on the east coast of Saudi Arabia (Khafji, Jubial, Dammam, and Salwa). Sediment samples were collected from both high tide and low tide zone. A total of 586 microplastic particles were collected from all the sites with an average particle size of 1.55 ± 0.94 mm. The majority of microplastic particles (77%) were less than 2 mm in size. Microplastic abundance ranged from 5.5 ± 1.55 to 21.2 ± 0.68 particle/kg (51.1 ± 14.71 to 152.8 ± 21.32 particle/m^2) in low tide region, and from 6.3 ± 4.05 to 16.5 ± 4.98 particle/kg (50.6 ± 31.21 to 204.5 ± 64.15 particle/m^2) in high tide region. The most dominant colors were transparent (34%) and blue (30%), while the fiber was the most common shape (96%). Polyethylene terephthalates were the common polymer type of fibers, while polyethylene and high-density polyethylene were common in fragments and filaments.

Keywords: microplastic, marine contamination, sediments, Arabian Gulf, microplastic contaminants

1. Introduction

The plastic production rate is increasing exponentially since the 1950s reaching more than 400 million tons per year in 2020 [1]. People favor plastic over other materials mainly because of its properties, such as durability, lightweight, and flexibility. A study done by Ritchie and Roser in 2018 estimated that the total plastic floating on the global ocean surface is about 269,000 [2]. Plastic polymers are often combined with an additive, which includes chemical compounds, metals, or persistent organic pollutants (POP) [3–5]. Some of these additives are significantly toxic to humans if ingested, inhaled, or even during dermal contact [6]. Plastic litter can degrade or break into smaller micro-size plastic via various routes, such as mechanical forces, UV light from the sun, biological degradation, oxidation, or hydrolysis [7–9].

Microplastics are defined as synthetic polymer particles, that is, less than 5 mm in diameter. The first identification of microplastic was done in 1972 by carpenter et al. in the Sargasso Sea in plankton net trawls [10]. Microplastic is introduced to the marine environment either as primary or secondary microplastic. Primary microplastics are plastic particles that are manufactured to be less than 5 mm in size. Secondary microplastics are created by the fragmentation of large plastic products into smaller particles [11]. Microplastic is usually introduced to the marine environment through wastewater, surface runoff, or fragmentation of plastic products in landfill and coastal areas [11].

Microplastic pollution is measured in water, sediments, and organisms. In marine environments, beaches are considered to be the reservoir of macroplastic and microplastic debris [12]. They receive plastic pollution from land and transport it to coastal water, and then open ocean. However, the fate of microplastic spatial distribution is uncertain and it depends on several factors: (1) chemical structure of microplastic, (2) seawater density, (3) weather, (4) microplastic additives, (5) polymer type, (6) ecological impact, and (7) fragmentation ability [13, 14]. Rivers are also considered a major source of microplastic contamination in the marine environment, and this is because they usually pass through several urban areas before discharging into the ocean [15].

Microplastics can interact with marine organisms mainly through ingestion due to their small size similar to organisms' natural food. Several studies were done to measure microplastic contamination and risk in microorganisms [10, 16, 17], fish and mammals [18–21], and birds [22]. It was found by scientists that biota exposed to microplastic will have negative health effects, such as decreasing food consumption [23], decrease in weight [24], growth rate [25], and fertility [26]. In the aquatic environment, bivalves are the most commonly used organism in the labs for exposure studies [27]. Also, because of their filter-feeding behavior, bivalves are used in several studies as a bioindicator for microplastic contamination [28].

2. Study area

Arabian Gulf is an important sea to the surrounding countries due to the existence of huge oil and gas reservoirs, supplying the countries with domestic water through desalination plants, and because of its richness in a variety of biological resources that supply the countries. Saudi Arabia has a coastline of around 800 km long on the Arabian Gulf starting from Khafji and ending in Salwa Bay [29]. The kingdom's territorial water covers an area of 27,050 km^2, which is more than 10% of the total Arabian Gulf area (240,000 km^2) [30]. Arabian Gulf is considered one of the most stressful environments for marine organisms due to its high salinity caused by the high temperate and low precipitation rate [29]. The gulf is a semi-enclosed sea surrounded by arid lands in the west and the Zagros mountains in the east and connected to the Indian Ocean through the Straits of Hormuz. The average depth of the gulf is 35 m and reaches a maximum depth of 100 m near the Straits of Hormuz [29]. The circulation in the Arabian Gulf is counterclockwise mainly driven by wind, and thermohaline [29]. A model of the gulf was created by Yousef Alosairi et al. [31] using a three-dimensional numerical model estuary, lake and coastal ocean model (ELCOM) which shows that the flushing time along the Arabian coast is more than 3 years, a slow process compared to other seas. The Arabian Gulf is surrounded by eight developing countries, including Saudi Arabia, Bahrain, Qatar, United Arab Emirates, Oman, Kuwait, Iraq, and Iran. Each of these countries has several cities and projects along the Arabian Gulf coastline, such as desalination plants, treated sewage disposal, nuclear plants, and many oil and gas production industries.

Only 17 published studies related to microplastic were done in the Arabian Gulf in which most of which are on Iranian coasts and the Strait of Hormuz. Therefore,

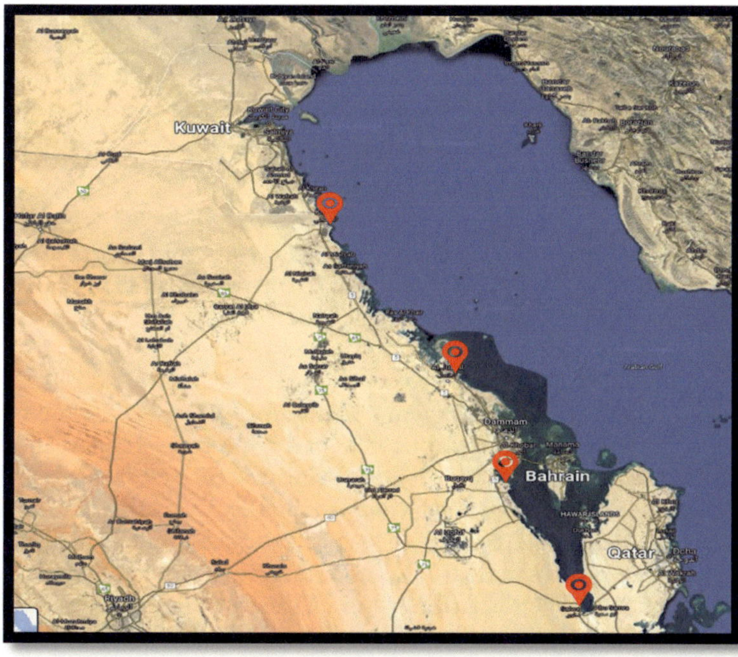

Figure 1.
Location of the four different beaches on Saudi Arabia's east coast.

the objective of this study is to characterize and compare microplastic contamination in the sediment at Saudi Arabia's east coast at four different beaches (Khafji, Jubail, Dammam, and Salwa Bay) (**Figure 1**). By measuring the following parameters:

- The microplastic abundance will be reported as particle/kg of sand, and particle/m^2.
- The microplastic size will include sizes from 5 mm down to 1 mm.
- The microplastic shape which usually classifies as fragment, foam, fiber, or pellet.
- Microplastic polymer types will be identified using ATR-FTIR.

3. Methodology

3.1 Sampling

To have a representative sample, samples were collected from beaches that were not developed and prepared for recreational purposes. The sand of recreational beaches usually is replaced or cleaned on regular basis. Therefore, it is recommended to sample from the original uncleaned sand. Moreover, low tide (LTZ) and high tide (HTZ) zones must exist and be identifiable in the sampling site to achieve our sampling objective. Northing and easting coordinates were recorded for each sample using global positioning system (GPS). Six samples were collected from each beach of which three samples were from the high tide zone (HTZ), and three from the low tide zone (LTZ). A wooden quadrat with 50× 50 cm was used for sampling with a small metal shovel. After removing the debris and other liters, the top 5 cm sediment within the quadrat was collected in a metal bucket, weighted, and covered with foil aluminum. Then, 2 kg of each sample was filtered through 5, and 0.5 mm mesh-size stainless steel sieves. The remaining sand in between the sieves was collected in labeled glass bottles and sent to the lab for the next steps.

3.2 Sample preparation

The density separation method was used to separate microplastic from the sand. Sodium Chloride salt (NaCl) was added gradually to a beaker filled with distilled water. The beaker was placed under a magnetic stirrer for continuous stirring. NaCl was added to the distilled water up to saturation point when the salt is not dissolving in the water anymore. To confirm the density of the brine, the weight of 1 L of brine was measured which was around 1.18 g/cm^3. Sand samples were extracted from the glass bottles into a stainless-steel container using a stainless-steel spoon and distilled water to make sure all the samples is collected. Brine was filtered through 100 μm mesh and then added to the containers using a volume equal to two to three times the sand volume to submerge the sample. Sand then was stirred for 5 minutes to allow light components like microplastics to float. The container then was covered with aluminum foil and kept for 24 hours to settle. After 24 hours, the brine in the samples along with the supernatant material was filtered through 100 μm stainless-steel mesh. The materials caught in the mesh were then washed using distilled water into a glass beaker. Then, the collected materials in the beaker were filtered using a 47 mm Whatman glass microfiber filter, ceramic funnel, and electric vacuum pump. Glass microfiber filter was then kept in a glass petri dish to dry and covered with aluminum foil.

3.3 Sample analysis

Glass filters are then analyzed under the stereo microscope "Olympus" with 450× magnification for any suspected microplastics. Suspected particles are collected using metal forceps and placed in a glass slide to be analyzed using Leica CME 1000× compound microscope. Particles are then identified as microplastic based on the following criteria [32–33]: 1. the absence of organic and cellular structures, 2. the color is homogenous, and 3. not segmented and evenly thick. After that, the color and shape of microplastics were recorded and a picture was taken of each sample with the scaled optical glass inserted in the microscope lens. Later with the help of a calibration slide, the length of each particle was measured using IC Measure software. Part of the identified microplastics was isolated in a small glass bottle and was analyzed for polymer types using Bruker ATR-FTIR. To cover all possible polymer types, samples with different characteristics (shape and color) were used. The resulting spectrums were matched with referenced polymer spectra using the library in OPUS-spectroscopy software. Statistical analysis was performed on the data using Microsoft excel. First, using the Kolmogorov–Smirnov test for normal distribution. Then, one-way analysis of variance (ANOVA) was applied for microplastic abundance, size, and color considering $p < 0.05$ as statistically significant.

4. Results

Microplastics were found in all 24 samples taken from the four different beaches Dammam, Jubail, Khafji, and Salwa (**Table 1**). A total of 586 particles were extracted

Area	Sample		Abundance		Size		Shape				Color		
DMMM		Total MP	Particle/kg	Particle/m^2	Ave Size	Fiber	Fragment	Filament	Transparent	Blue	Black	Red	Green
	DMMM-1 LT	42	22.1	141	1.7	39	2	1	18	9	2	12	1
	DMMM-2 LT	43	20.5	134	1.47	37	5	1	15	14	9	4	1
	DMMM-3 LT	44	21.0	183	1.62	44	0	0	24	12	3	4	1
	DMMM-1 HT	17	8.5	62	1.75	11	6	0	5	7	3	2	0
	DMMM-2 HT	35	16.7	124	1.51	34	0	1	11	11	7	6	0
	DMMM-3 HT	24	12.0	87	1.3	24	0	0	6	6	8	3	1
JUBL	JUBL-1 LT	14	7.0	66	1.36	14	0	0	7	4	1	2	0
	JUBL-2 LT	7	3.3	31	1.21	7	0	0	0	2	3	2	0
	JUBL-3 LT	14	6.1	57	1.63	14	0	0	3	7	3	0	1
	JUBL-1 HT	3	1.5	13	2.55	2	1	0	2	0	0	0	1
	JUBL-2 HT	24	11.4	90	1.68	24	0	0	8	6	4	4	2
	JUBL-3 HT	12	6.0	49	1.62	10	2	0	4	3	1	2	2
KHFJ	KHFJ-1 LT	39	17.7	194	1.52	38	1	0	14	8	7	7	3
	KHFJ-2 LT	20	9.1	96	1.16	20	0	0	6	8	6	0	0
	KHFJ-3 LT	28	12.2	139	1.52	28	0	0	4	9	6	6	3
	KHFJ-1 HT	15	7.1	77	1.32	15	0	0	3	7	2	1	2
	KHFJ-2 HT	20	9.1	76	1.27	20	0	0	4	2	3	9	2
	KHFJ-3 HT	21	10.0	84	1.13	21	0	0	7	6	5	2	1

Area	Sample		Abundance		Size		Shape				Color		
SALW	SALW-1 LT	21	10.5	123	1.7	18	2	1	8	8	2	2	1
	SALW-2 LT	12	5.5	66	1.8	11	1	0	6	1	1	4	0
	SALW-3 LT	21	9.1	120	1.64	21	0	0	8	7	5	1	0
	SALW-1 HT	48	20.9	265	1.85	48	0	0	21	14	6	7	0
	SALW-2 HT	42	19.1	232	1.53	42	0	0	8	20	7	7	0
	SALW-3 HT	20	9.5	116	1.44	19	1	0	7	7	3	3	0

Table 1.
Detailed results of all of the samples analyzed for the microplastic study.

with an average size of 1.55 ± 0.94 SD mm and an overall average abundance of 11.5 ± 5.02 SD particle/kg (109.4 ± 50.26 SD particle/m^2). The highest abundance among the low tide is found in the Dammam sample with an average of 21.2 ± 0.68 SD particle/kg (152.8 ± 21.3 SD particle/m^2), while the lowest is in the Jubail sample with an average of 5.5 ± 1.56 SD particle/kg (51.1 ± 14.71 SD particle/m^2). On the other hand, Salwa has the highest abundance among the high tide with 16.5 ± 4.98 SD particle/kg (204.5 ± 64.15 SD particle/m^2) and Jubil is the least with 6.3 ± 4.06 SD particle/kg (50.6 ± 31.21 SD particle/m^2) (**Figures 2** and **3**). No significant differences were reported between the overall low tide average abundance (12 ± 5.94 particle/kg, 112.5 ± 40.08 SD particle/m^2) and overall high tide average abundance (11 ± 3.85 particle/kg, 106.3 ± 58.54 SD particle/m^2). However, when low tide and high tide average microplastic abundance were compared for each site separately, discrepancies were observed in all of the beaches except the Jubail sample. The low tide abundance average was higher than the high tide in Dammam and Khafji, but the opposite was reported in Salwa.

Average particle size was showing no significant variation between both low tide (1.53 mm) and high tide (1.58 mm). Microplastics with a size range from 1 to 2 mm were the most common size with 44%, followed by particles that were less than 1 mm in size (33%), then the range from 2 to 3 mm (14%). Therefore, only 9% of the collected particles were bigger than 3 mm (**Figure 4**).

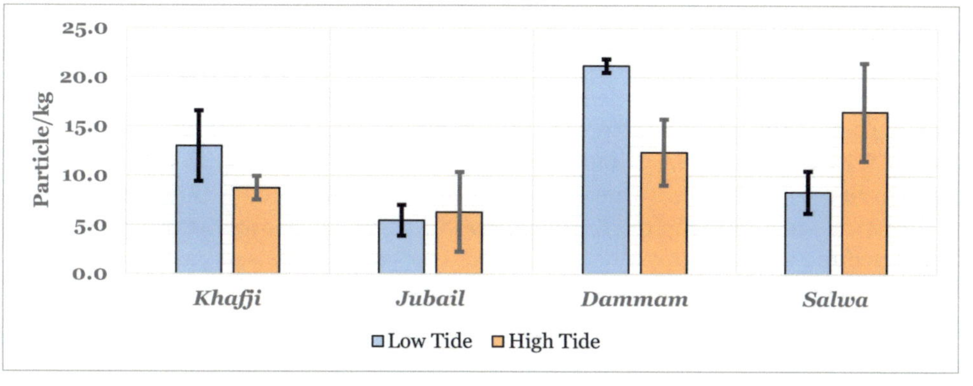

Figure 2.
Average microplastic abundance (particle/kg).

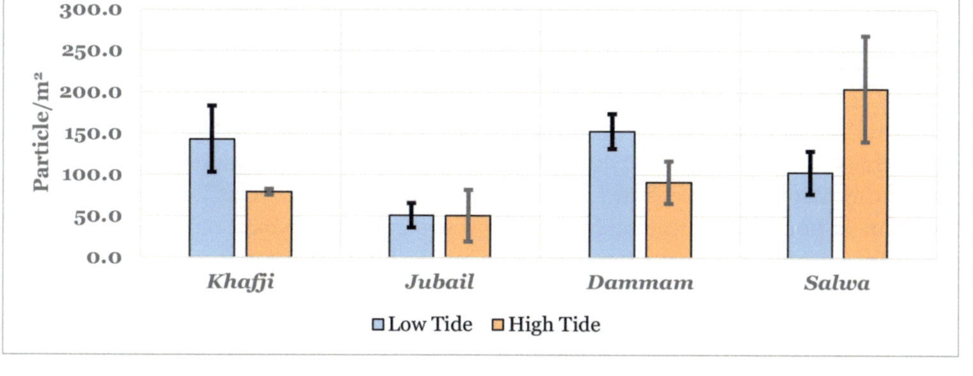

Figure 3.
Average microplastic abundance (particle/m^2).

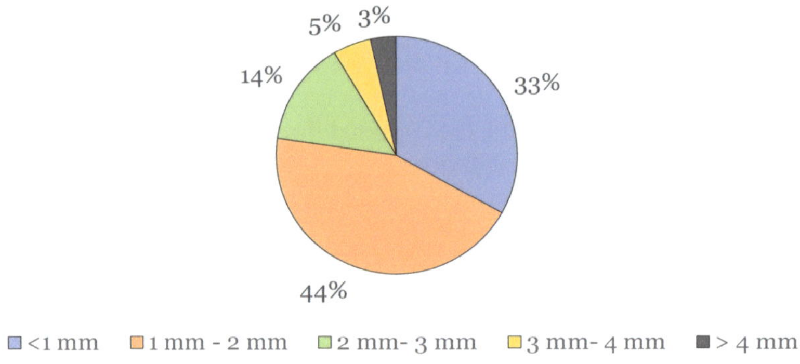

Figure 4.
Microplastic size distribution.

Regarding microplastic types, 96% of the collected particles were fibers leaving only 4% as fragments and filaments. Fragments and filaments were found in both low-tide and high-tide samples.

Different colors of microplastic particles were observed in all the samples. The majority is for transparent (34%) followed by blue (30%), black (17%), red (15%), and green (4%).

Using ATR-FTIR analysis, four different polymers as polyethylene terephthalate (PET), polyethene (PE), high-density polyethylene (HDPE), and polypropylene (PP) were identified. Fibers with different colors (transparent, blue, and red) were found to be PET. PE and HDPE were found in fragments and filaments with blue, green, and transparent colors while only one white fragment was found in PP (**Figure 5**).

Statistical analysis using the Kolmogorov–Smirnov test shows that microplastic abundance and size are normally distributed. One-way analysis of variance (ANOVA) shows no significant differences between microplastic abundance, size, and color considering $p < 0.05$ as statistically significant.

5. Discussion

Numerous studies were conducted around the world to characterize microplastic abundance in water, sediment, and biota. However, still the methodology and reporting unit is not yet standardized. Different sampling procedures for sediment can be found in the articles. For example, sample quadrat varies in the studies from 1 × 1 m to 0.3 × 0.3 m [34, 35], and the most common quadrat is 0.5 × 0.5 m. This is the case with sampling thickness in which some studies collect the sample from 5 cm [36] while others take only the top 1 cm [33, 36, 37]. Also, density separation fluid density varies from 1.2 g/ml for NaCl to 1.8 g/ml for NaI. In this study, NaCl was used since it is the most popular choice, easy to use, and low cost. Moreover, reporting units in the published papers are different such as particle/kg, particle/m^2, mg/g sediment, mg/m^2, and particle/0.0125m^3. Weight in some studies is presented as dry sediment weight, while in others is a wet sediment weight. Therefore, comparison between the results is not a straightforward process. Multiple published articles and review papers summarized the methodologies and results of many microplastic studies done around the world [38–40].

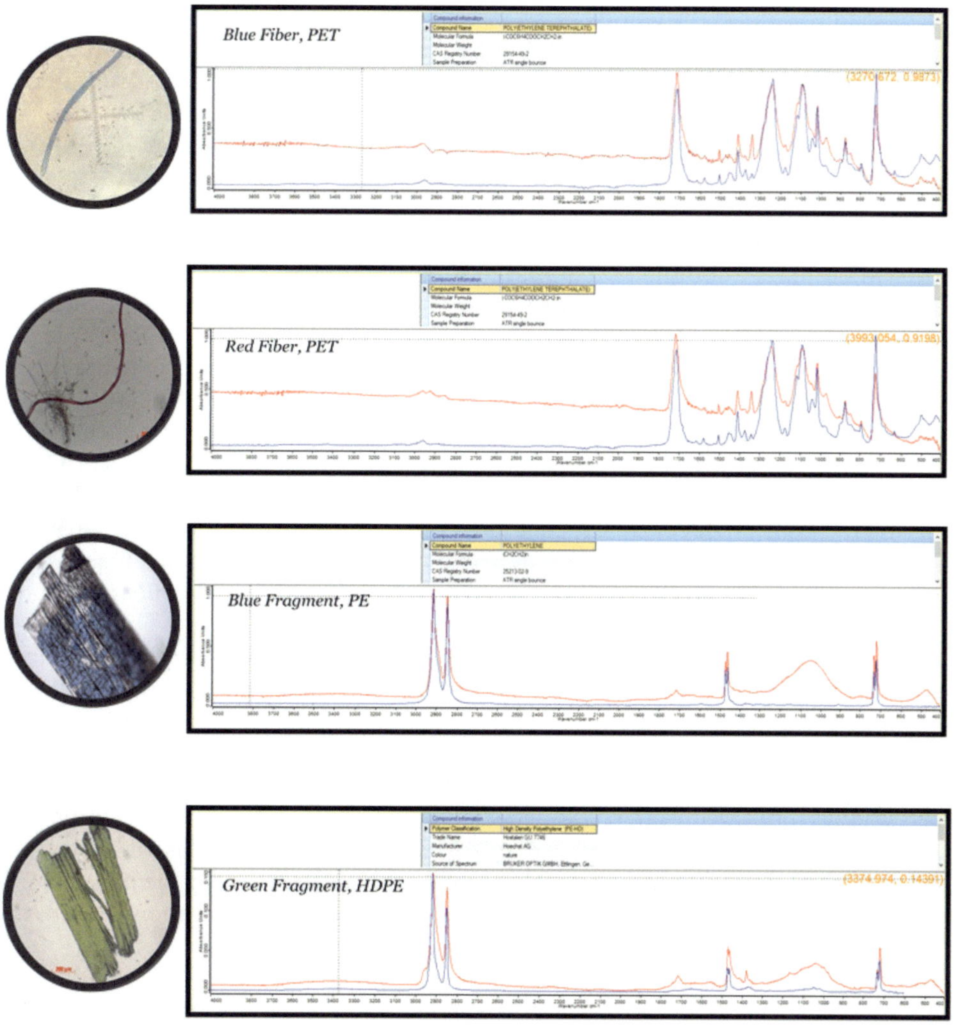

Figure 5.
Spectrum for different shapes and colors of microplastic.

Microplastic abundance per area was estimated in this study since part of the studies done in the Arabian Gulf described microplastic abundance as particle/m^2 [33, 41]. To calculate the abundance in particle/m^2, equal vertical distribution along the 5 cm depth must be assumed. Also, it is important to clarify that the reporting unit is a particle of microplastic in an area of 1 m^2 and at 1 cm depth. The results showed an average abundance of 112.5 particles/m^2 for the low tide zone and 106.3 particles/m^2 for the high tide zone.

Comparing microplastic abundance on the east coast of Saudi Arabia against worldwide countries shows that the Saudi east coast of the Arabian Gulf is among the least polluted coasts (5.5 to 21.2 particles/kg, 50.6 to 204.5 particles/m^2). Multiple studies with similar microplastic extraction procedure done on the southwestern side of the USA coast and at several European beaches [42, 43] shows a higher abundance of microplastic [42–44]. Same with Asian countries where microplastic abundance tends to be higher than our results [12, 45, 46]. **Table 2** summarizes some results of various studies around the world taking into consideration similar methodologies and reporting units. Higher abundance in smaller particle sizes is also observed in most of the studies. Black and blue were the most common colors in these articles, and fiber is the common shape.

With regards to the gulf countries, a review paper by Saif Uddin et al. [38] summarizes all the studies carried out in the Arabian Gulf in terms of the type of samples, sampling methodology, and results. From the Iranian coast to the Oman sea through the Strait of Hormuz to Bander Abbas city beaches, studies show greater microplastic abundance compared to the Saudi east coast [36, 41, 47, 48]. This might be due to the geology of the area and the counterclockwise movement of the seawater [29]. In these studies, density separation was done using brine density higher than NaCl which led to extracting more microplastic particles. As stated by Naji et al. [49] in their study that around 74% of microplastic was extracted using NaCl, while the remaining 26% was recovered by NaI. However, the studies done in Qatar by Abayomi [50] and UAE by Aslam [33] used the KI solution for density separation and it shows much lower microplastic abundance than that of the Iranian coast. The results of both studies are comparable with this study's numbers and that gives more confidence in our methodology. This was also supported by the similarity in having fiber, blue, and PE as the most common properties of the collected microplastic.

Despite that the average of low tide and high tide abundance is very close to each other; it may vary a lot when compared site by site. The difference is observed in all the sites except Jubial which has the lowest abundance among all the sites. Low tide shows higher abundance in Dammam and Khafji samples while it was less in Salwa. Therefore, there is no trend between the high tide and low tide abundance which was also observed [47]. It was expected to have higher microplastic abundance in Dammam, Jubail, and Khafji since they are both crowded and industrial cities. However, Jubail shows a very low number which was unexpected for the biggest industrial city in the eastern region. Several reasons might cause these low numbers which are as follow:

1. The geology and location of Jubail city where the counterclockwise current is blocked by Abu Ali Island in the north.

2. The effective recycling of the wastewater, which is treated and reused for irrigation instead of discharging it in the sea.

3. The frequent cleaning of the developed beach will reduce the amount of plastic runoff to the seawater.

On the other hand, Salwa which is considered a rural city showed high microplastic abundance and this also can be explained by the following reasons:

1. Although Salwa has a very low population, the sea there is surrounded by land from three directions which makes it a trap for floating contamination.

2. Multiple resorts in Qatar side in front of Salwa beach.

3. Salwa beach is considered a destination for tourism to practice fishing and camping.

Since Salwa is located near the border of Qatar, it will be useful to compare its results with Abayomi's results [50]. Umm Bab sampling site is the closest one to Salwa beach, and it shows a microplastic abundance of 8.3 particles/kg which is similar to our results for Salwa average low tide (8.4 particles/kg) (**Figure 4**).

Microplastic with a size ≤2 mm was the most common size among the samples (77%). It is expected to have even smaller (<0.5 mm) microplastics that already passed the 0.5 mm stainless steel sieve. The smaller the microplastic particles are, the more vulnerable they became to a smaller organism. Based on a study done by Naji et al. [49], microplastics with sizes ranging from 0.02 to 1.68 mm were the most common size found in cyclopoids, shrimps, and zoea. Having microplastic induced in the organisms at the base of the food web could cause biomagnification in the bigger organisms that feed on them.

As mentioned before, around 96% of the collected particles were fiber. This result is similar to all the studies done in the Arabian Gulf and most worldwide studies (**Tables 2** and **3**). This could be due to disposing sewage rich in synthetic fiber released from textiles, or clothes into the sea [6, 57]. Degraded fishing ropes and nets are considered a source of microplastic fiber in seawater. Treatment of wastewater before disposing to the sea may remove more than 80% of the microplastic [58]. Unfortunately, some countries dispose of the wastewater without primary treatment [59] which may increase the microplastic fiber in the oceans. Microplastic fibers were found to be one of the most shapes to be ingested by zooplankton [33, 60, 61]. More studies also showed that microplastic fiber is the most ingested by turtles, fish larvae, and Mesoplodon mirus [62–64].

Transparent and blue colors like most of the studies in the area were the most common colors followed by black color (**Figure 6**). No significant change in color distribution was found among the sampling sites. The importance of quantifying microplastic color was described by Shaw et al. [65] research which shows that colored microplastic is more likely to be ingested by organisms as prey, and the white, transparent, blue, and black colors were the most commonly found [65]. During microscope analysis, some particles were found with a partially faded color. This indicates that some of the originally colored microplastic particles turned into colorless particles and that could be because of the physical or chemical reactions between the particles and the surroundings. The changing of color was also mentioned by Chen et al. [66] and by Wibowo et al. [67] which show that time is playing a role in changing the microplastic color and size.

Attenuated total reflectance fourier transform infrared spectroscopy (ATR-FTIR) analysis shows that different colors of microplastic fibers had similar polymer types which are polyethylene terephthalate (PET). Since fiber was the most common shape (96%), then this suggests that PET is the most common polymer type in the collected samples. This result is consistent with the studies done in the Arabian Gulf, and most of the studies around the world. PET is commonly used in making ropes and drinking bottles which are heavily used in KSA specifically for water bottles. Fragments and filaments particle were identified as polyethylene (PE) and high-density polyethylene (HDPE). These polymer products are also commonly used in our daily life, such as bags and shampoo bottles. Only one particle was identified as polypropylene (PP) which was also reported in Qatar by Abayomi et al. [50]. This suggests that most of the microplastic contamination in Saudi east beach is coming from the daily used plastic product which reaches the sea through wastewater effluent, direct dumping of plastic into the sea, or run-off.

Country	Location	Abundance		Average Particle Size	Most Common			Reference
		Particle/kg	Particle/m2		Shape	Color	Polymer	
USA	Southeastern Beach	43–306 (D. W.)	—	—	Fiber	—	PET	[44]
	Mobile bay, Alabama	50.6 ± 9.96	—	2.5 mm	Fragment		PE & PP	[51]
	Kamilo/Kahuku Beach, Hawai'i	—	1143/1774	2–4 mm	Fragment	White	PE	[52]
Italy	Sicily	160 ± 31 (D. W.)	—	1.32 mm	Fiber	Blue/Black	—	[42]
	Tyrrhenian Sea	—	151–678.7	—	Filament	—	—	[53]
Spain	Denia	156 ± 29 (D. W.)	—	1.96 mm	Fiber	Blue/Black	—	[42]
	Lambra/Famara/Las Caneras	—	430.9–1656	—	Fragment	—	—	[54]
France	Normandy	143 ± 13 (D. W.)	—	1.28 mm	Fiber	Blue/Black	—	[42]
UK	Scapa Flow Orkney	730–2300 (D. W.)	—	—	Fiber	Blue	PTFE	[43]
Netherlands	Rottumeroog	124 ± 27 (D. W.)	—	1.28 mm	Fiber	Blue/Black	—	[42]
Denmark	Fyns Hoved	164 ± 21 (D. W.)	—	1.26 mm	Fiber	Blue/Black	—	[42]
Russian	Baltic Coast	0.2–36.2 (D. W.)	5560	—	Foam	—	—	[55]
China	Guangdong Province	—	6675	0.315–1 mm	Foam		PS	[12]
South Korea	Nakdong River Estuary	—	27,606	—	Foam	—	—	[45]
Maldives	Coral Island	—	647 ± 720	—	Foam	—	PE	[46]
South Africa	Indian Ocean	—	688.9 ± 348.2 & 3308 ± 1449		Fiber	Blue/Black	PS	[56]

Table 2.
Examples of microplastic studies around the world.

Country	Location	Abundance		Density Separation Brine	Most Common			Reference
		Particle/kg	Particle/m^2		Shape	Color	Polymer	
Iran	Strait of Hormuz	2 ± 1 to 1258 ± 291 (D. W.)	—	NaCl	Fiber		PET & Nylon	[36]
	Oman Sea	138.3 ± 4.5 to 930.3 ± 49 (D. W.)	—	NaCl & ZnCl2	Fiber & Fragment	White	PE	[48]
	Bandar Abbas	—	3252 ± 2766	—	Foam	—	PS & PET	[41]
	Bandar Abbas	36 ± 7.2 to 125 ± 25 (D. W.)	—	NaCl & NaI	Fiber	—	PE & PET	[47]
	Khor-e-Khoran Mangrove	19.5 to 34.5 (D. W.)	—	NaCl & NaI	Fiber	Black	—	[49]
Qatar	Qatar Coasts	—	36 to 228	KI	Fiber	Blue	PE & PP	[50]
UAE	Dubai Coast	59.71 (D. W.)	165	KI	Fiber	Blue	PE	[33]
Kuwait	Kuwait Coasts	—	—	—	Filament & Fragment	Blue	PP & PE	[35]

Table 3.
Microplastic studies in the Arabian Gulf (sediment).

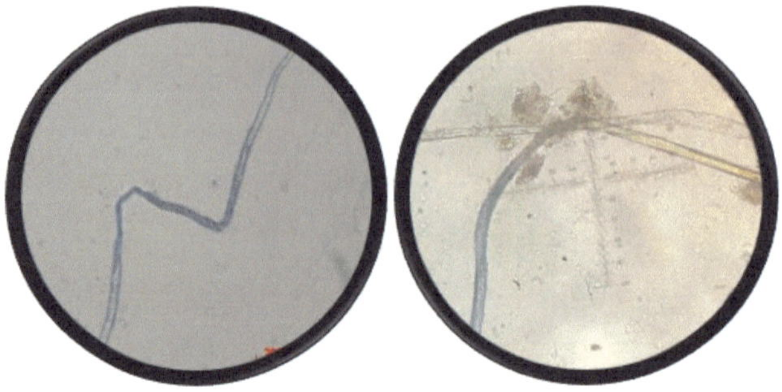

Figure 6.
Microplastic color fading and changing.

6. Conclusion and recommendation

This study is the first study to measure microplastic pollution in the sediment of the eastern coast of the kingdom of Saudi Arabia. The common size of microplastic particles measured (77%) was less than 2 mm in size. Microplastic abundance within the study sites ranged from 5.5 to 21.2 particles/kg (51.1 to 152.8) in the low tide region, and from 6.3 to 16.5 particles/kg (50.6 to 204.5 particles/m^2) in the high tide region. The main goal of this study is to provide a baseline and glimpse of the quantity and identity of microplastic in our selected sites, however, those values consider to be a baseline for a coast length of 800 km. The future researcher will have some expectations of microplastic distribution along this coastline. Several recommendations can be suggested to whom is interested in curry microplastic research on Saudi Arabia's eastern coast:

- Identify the effect of sampling time in microplastic quantity by collecting the sample from a single location at a different time interval.
- Measure microplastic contamination in both water and sediment in the same area to find the relationship.
- Identify an organism to be used as a bioindicator in our area, such as Clam in China [28].
- Use multiple brine solutions with different densities to find the difference between the quantity and type of microplastic extracted from each one.
- Measure microplastic abundance in sediment with respect to depth
- Provide the results with a different unit to allow comparison with different studies.
- Measure microplastic in a protected area (mangroves and islands) which is important to preserve the area and the organisms living there.
- Digestion of organic material by acids can make microplastic identification easier, however, it can affect color and shape identification.
- Filtering the supernatant using a fine mesh size stainless steel (300 μm) can ease the filtration of supernatant later in the glass filter paper.

Acknowledgements

The authors would like to thank King Fahd University of Petroleum and Minerals (KFUPM) for their support by providing all the materials and equipment to conduct the study.

Author details

Hamza Jawad Al-Shaikh Ali[1], Assad Ahmed Al-Thukair[2]*, Abdurahiman Kambrath Pulikkoden[3] and Basheer Chanbasha[4]

1 Environmental Science Program, Department of Geoscience, Collage of Petroleum Engineering and Geoscience, King Fahad University for Petroleum and Minerals (KFUPM), Dhahran, Saudi Arabia

2 Department of Bioengineering, Collage of Chemical and Materials, King Fahad University for Petroleum and Minerals (KFUPM), Dhahran, Saudi Arabia

3 Center for Environment and Marine Studies, Research Institute, King Fahad University for Petroleum and Minerals (KFUPM), Dhahran, Saudi Arabia

4 Department of Chemistry, College of Chemical and Materials, King Fahad University for Petroleum and Minerals (KFUPM), Dhahran, Saudi Arabia

*Address all correspondence to: thukair@kfupm.edu.sa

References

[1] Isangedighi IA, David GS, Obot OI. Plastic waste in the aquatic environment: Impact and management. Environment Science and Policy for Sustainable Development. 2018;**2**:2

[2] Ritchie H, Roser M. Plastic pollution. September 2018. [Online]. Available from: https://ourworldindata.org/plastic-pollution

[3] Teuten EL, Saquing JM, Knappe DR, Barlaz MS, Rowlan J, et al. Transport and release of chemicals from plastics to the environment and to wildlife. Philosophical Transactions of The Royal Society B. 2009;**364**:2027-2045

[4] Antunes JC, Frias J, Micaelo A, Paula S. Resin pellets from beaches of the Portuguese coast and adsorbed persistent organic pollutants. Estuarine Coastal and Shelf Science. 2013;**130**:62-69

[5] Massos A, Turner A. Cadmium, lead and bromine in beached microplastics. Environmental Pollution. 2017;**227**:139-145

[6] UNEP. Marine Plastic Debris and Microplastics. Nairobi: United Nations Environment Programme; 2016

[7] Andrady AL. Microplastic in The Marine Environment. Marine Pollution Bulletin. 2011;**62**:1596-1605

[8] Browne MA, Galloway T, Thompson R. Microplastic—An emerging contaminant of potential concern? Integrated Environmental Assessment and Management. 2007;**3**:559-561

[9] Barenes DK, Galgani F, Thompson RC, Barlaz M. Accumulation and fragmentation of plastic debris in global environment. Philosophical Transactions of The Royal Society B Biological Sciences. 2009;**364**

[10] Carpenter EJ, Smith KL Jr. Plastics on the sargasso sea surface. ScienceMag. 1972;**175**:1240-1241

[11] GESAMP. Sources, fate and effects of microplastics in the marine environment: A global assessment. In: Kershaw PJ, editor. IMO/FAO/UNESCO-IOC/UNIDO/WMO/IAEA/UN/UNEP/UNDP. Joint Group of Experts on the Scientific Aspects of Marine Environmetal Protection. 2015. p. 96

[12] Fok L, Cheung PK, Tang G, Li WC. Size distribution of stranded small plastic debris on the coast of Guangdong, South China. Environmental Pollution. 2017;**220**:407-412

[13] Andrady AL. The plastic in microplastics: A review. Marine Pollution Bulletin. 2017;**119**:12-22

[14] Arias-Andres M, Klumper U, Rojas-Jimenez K, Grossart H-P. Microplastic pollution increases gene exchange in aquatic ecosystems. Environmental Pollution. 2018;**237**:253-261

[15] Sighicelli M, Pietrelli L, Lecca F, Lannili V, Falconieri M, Coscia L, et al. Microplastic pollution in the surface waters of Italian Subalpine Lakes. Environmental Pollution. 2018;**236**:645-651

[16] Cole M, Lindeque P, Fileman E, Halsband C, Goodhead R, Moger J, et al. Microplastic ingestion by Zooplankton. Environmental Science & Technology. 2013;**47**:6646-6655

[17] Bhattacharya P, Lin S, Turner JP, Chun Ke P. Physical adsorption of charged plastic nanoparticles affects algal photosynthesis. The Journal of Physical Chemistry C. 2010;**114**:16556-16561

[18] Lusher AL, McHugh M, Thompson RC. Occurrence of microplastics in the gastrointestinal tract of pelagic and demersal fish from the English channel. Marine Pollution Bulletin. 2013;**67**:94-99

[19] Baalkhuyur FM, Bin Dohaish E-JA, Elhalwagy ME, Alikunhi NM, AlSuwailem AM, Rostad A, et al. Microplastic in the gastrointestinal tract of fishes along the Saudi Arabian Red Sea Coast. Marine Pollution Bulletin. 2018;**67**:94-99

[20] Abbasi S, Soltani N, Keshavarzi B, Moore F, Turner A, Hassanaghaei M. Microplastics in different tissues of fish and prawn from the Musa Estuary, Persian Gulf. Chemosphere. 2018;**205**:80-87

[21] Boerger CM, Lattin GL, Moore SL, Moore CJ. Plastic ingestion by planktivorous fishes in the North Pacific Central Gyre. Marine Pollution Bulletin. 2010;**60**:2275-2278

[22] Bond AL, Provencher JF, Elliot RD, Ryan PC, Rowe S, Jones IL, et al. Ingestion of plastic marine debris by common and thick-billed Murresin the northwestern Atlantic from 1985 to 2012. Marine Pollution Bulletin. 2013;**77**:192-195

[23] Besseling E, Wegner A, Foekema EM, Heuvel-Greve MVD, Koelmans A. Effects of microplastic on fitness and PCB bioaccumulation by the lugworm Arenicola marina (L.). Environmental Science & Technology. 2013;**47**:593-600

[24] Wright S, Rowe D, Thompson RC, Galloway TS. Microplastic ingestion decreases energy reserves in marine worms. Current Biology. 2013;**23**:R1031-R1033

[25] Huerta E, Gertsen H, Gooren H, Peters P, Salanki T, Van Der Ploeg MJ, et al. Microplastics in the terrestrial ecosystem: Implications for lumbricus terrestris (oligochaeta, lumbricidae). Environmental Science & Technology. 2016;**50**:2685-2691

[26] Sussarellu R, Suquet M, Thomas Y, Lambert C, Fabioux C, Pernet EJ, et al. Oyster reproduction is affected by exposure to polystyrene microplastics. Proceedings of The National Academy of Sciences. 2016;**113**:2430-2435

[27] Lusher A, Hollman P, Mendoza-Hill J. Microplastics in fisheries and aquaculture, food and agriculture organization of the United Nation. 2017

[28] Su L, Cai H, Kolandhasmy P, Wu C, Chelsea RM. Using the Asian clam as an indicator of microplastic pollution in freshwater ecosystems. Environmental Pollution. 2018;**234**:347-355

[29] Krishnakumar PK, Ashraf M, Sheppard C, M. J. D. Chapter 2: The physical environment of the Western Arabian Gulf. In: Ecosystems and biodiversity of the Arabian Gulf-Saudi Arabian Waters, Dhahran, Saudi Aramco and KFUPM. 2019

[30] Abuzinada AH, Barth H-J, Krupp F, Boer B, Al Abdessalaam TZ. Protecting the Gulf's Marine Ecosystem from Pollution. Switzerland: Birkhauser; 2008

[31] Alosairi Y, Imberger J, Falconer RA. Mixing and flushing in the Persian Gulf (Arabian Gulf). Journal of Geophysical Research. 2011;**116**:1-14

[32] Norén F. Small Plastic Particles in Coastal Swedish Waters. N-Reserarch. Sweedn-Lysekil; 2007

[33] Aslam H, Ali T, Mortula MM, Attaelmanan AG. Evaluation of microplastics in beach sediments along the coast of Dubia, UAE. Marine Pollution Bulletin. 2020;**150**:110739

[34] Karthik R, Robin R, Purvaja R, Ganguly D, Anadavelu I, Raghuraman R, et al. Microplastics along the beaches of Southeast coast of India. Science of the Total Environment. 2018;**645**:1388-1399

[35] Saeed T, Al-Jandal N, Al-Mutairi A, Taqi H. Microplastic in Kuwait marine environment: results of first survey. Marine Pollution Bulletin. 2020;**152**:110880

[36] Naji A, Esmaili Z, Khan FR. Plastic debris and microplastics along the beaches of the strait of Hormuz, Persian Gulf. Marine Pollution Bulletin. 2017;**114**:1057-1062

[37] Enrique De-la-Torre G, Dioses-Salinas DC, Marlith Castro J, Antay R, Fernandez NY, Espinoza-Morriboron D. Abundance and distribution of microplastic on sandy beaches of Lime, Peru. Marine Pollution Bulletin. 2020;**151**:110877

[38] Uddin S, Fowler SW, Saeed T. Microplastic particles in the Persian/ Arabian Gulf – A review on sampling. Marine Pollution Bulletin. 2020;**154**:111100

[39] Rezania S, Park J, Md Din MF, Taib SM, Talaiekhozani A, Yadav KK, et al. Microplastics pollution in different aquatic environments and biota: A review of recent studies. Marine Pollution Bulletin. 2018;**133**:191-208

[40] Shahul Hamid F, Agamuthu P. Worldwide distribution and abundance of microplastic: How dire is the situation? Waste Managment & Research. 2018;**36**:873-897

[41] Nabizadeh R, Sajadi M, Rastkari N, Yaghmaeian K. Microplastic pollution on the Persian Gulf shoreline A Case Study of Bandar Abbas City, Hormozgan Province, Iran. Marine Pollution Bulletin. 2019;**145**:536-546

[42] Lots FA, Behrns P, Vijver MG, Horton AA, Bosker T. A large-scale investigation of microplastic contamination: Abundance and characteristics of microplastics in European beach sedimen. Marine Pollution Bulletin. 2017;**123**:219-226

[43] Blumenroder J, Sechet P, Kakkonen JE, Hatl MG. Microplastic contamination of intertidal sediments of Scapa Flow, Orkney: A first assessment. Marine Pollution Bulletin. 2017;**124**:112-120

[44] Yu X, Ladewig S, Bao S, Toline CA, Whitmire S, Chow AT. Occurrence and distribution of microplastics at selected coastal sitesalong the Southeastern United States. Science of the Total Environment. 2018;**613-614**:298-305

[45] Lee J, Hong S, Song YK, Hong SH, Jang YC, Jang M, et al. Relationships among the abundances of plastic debris in different sizeclasses on beaches in South Korea. Marine Pollution Bulletin. 2013;**77**:349-354

[46] Imhof HK, Sigl R, Brauer E, Feyl S, Giesmann P, Klink S, et al. Spatial and temporal variation of macro-, meso- and microplastic abundance on a remote coral island of the Maldives, Indian Ocean. Marine Pollution Bulletin. 2017;**116**:340-347

[47] Naji A, Esmaili Z, Mason SA, Vethaak D. The occurrence of microplastic contamination in littoral sediments of The Persian Gulf, Iran. Environmental Science and Pollution Research. 2017;**24**:20459-20468

[48] Kor K, Ghazilou A, Ershadifar H. Microplastic pollution in the littoral sediments of the Northern Part of the Oman Sea. Marine Pollution Bulletin. 2020;**155**:111166

[49] Naji A, Nuri M, Amiri P, Niyogi S. Small microplastic particles (S-MPPs) in sediments of mangrove ecosystem. Marine Pollution Bulletin. 2019;**146**:305-311

[50] Abayomi OA, Range P, Al-Gouti MA, Obbard JP, Almeer SH, Ben-Hamadou R. Microplastics in coastal environments of The Arabian Gulf. Marine Pollution Bulletin. 2017;**124**:181-188

[51] Wessel CC, Lockridge GR, Battiste D, Cebrian J. Abundance and characteristics of microplastics in beach sediments insights into microplastic accumulation in northern Gulf of Mexico estuaries. Marine Pollution Bulletin. 2016;**109**:178-183

[52] Young AM, Elliott JA. Characterization of microplastic and mesoplastic debris in sediments from Kamilo Beach and Kahuku Beach, Hawai'i. Marine Pollution Bulletin. 2016;**113**:477-482

[53] Faselli P, Blaskovic A, Bernardi G, Romeo T, Cizmek H, Andaloro F, et al. Plastic litter in sediments from a marine area likely to become protected (Aeolian Archipelago's islands, Tyrrhenian Sea). Marine Pollution Bulletin. 2016;**113**:526-529

[54] Herrera A, Asensio M, Martinez I, Santana A, Packard T, Gomez M. Microplastic and tar pollution on three Canary Islands beaches: An annualstudy. Marine Pollution Bulletin. 2016;**129**:494-502

[55] Esiukova E. Plastic pollution on the Baltic beaches of Kaliningrad region, Russia. Marine Pollution Bulletin. 2017;**114**:1072-1080

[56] Nel H, Froneman P. A quantitative analysis of microplastic pollution along the south-easterncoastline of South Africa. Marine Pollution Bulletin. 2015;**101**:274-279

[57] Rebelein A, Int-Veen I, Kammann U, Scharsach JP. Microplastic fibers - Underestimated threat to aquatic organisms? Science of the Total Environment. 2021;**777**:146045

[58] Shen M, Zhang Y, Almatarfi E, Hu T, Zhou C, Song B, et al. Efficient removal of microplastics from wastewater by an electrocoagulation process. Chemical Engineering Journal. 2022;**428**:131161

[59] Corcoran E, Nellemann C, Baker E, Bos R, Osborn D, Savelli H. Sick water. In: The Central Role of Wastewater Managment in Sustainable Development. United Nation Environment Programm; 2010

[60] Amin RM, Sohaimi ES, Anuar ST, Bachok Z. Microplastic ingestion by zooplankton in Terengganu coastal waters, southern South China Sea. Marine Pollution Bulletin. 2020;**150**:110616

[61] Desforges J-PW, Galbraith M, Dangerfield N, Ross PS. Widespread distribution of microplastics in subsurface seawaterin the NE Pacific Ocean. Marine Pollution Bulletin. 2014;**79**:94-99

[62] Duncan EM, Broderick AC, Fuller WJ, Galloway TS, Godfrey MH, Hamann M, et al. Microplastic ingestion ubiquitous in marine turtles. Global Change Biology. 2018;**25**:744-752

[63] Steer M, Cole M, Thompson RC, Lindeque PK. Microplastic ingestion infish larvae in the western English channel. Environmental Pollution. 2017;**226**:250-259

[64] Lusher AL, Hernandez-Milian G, O'Brien J, Berrow S, O'Connor I, Officer R. Microplastic and macroplastic ingestion by a deep diving, oceaniccetacean: The true's beaked whalemesoplodon mirus. Environmental Pollution. 2015;**199**:185-191

[65] Shaw DG, Day RH. Colour- and form-dependent loss of plastic micro-debris from the North Pacific Ocean. Marine Pollution. 1994;**28**:39-43

[66] Chen Q, Zhang H, Allgeier A, Zhou Q, Ouelle JD, Crawford SE, et al. Marine microplastics bound dioxin-like chemicals: Model explanation and risk assessment. Journal of Hazardous Materials. 2019;**364**:82-90

[67] Wibowo YG, Rosanti D. Microplastic in marine environment and its impact. Sainmatika. 2019;**16**:81-87

Chapter 10

Micro(Nano)Plastics as Carriers of Toxic Agents and Their Impact on Human Health

Karla Lizzeth Morales-Cano, Leandro Hermida-Castellanos, Carlos M. Adame-Adame, Luis Alberto Peralta Peláez and Carolina Peña-Montes

Abstract

Currently, globally the demand and production of plastic items are increasing exorbitantly, generating a large amount of waste, and polluting the ecosystem, a site in which degradation processes are triggered, which give rise to smaller particles such as micro(nano)plastics (MNP). Continuous human exposure to these particles generates negative alterations in the host's health. Three routes of MNP exposure or contact have been established: inhalation, ingestion of particles, and dermal absorption. Recently, it has been pointed out that microplastics (MP) can even be found in the human placenta. This chapter aims to compile and provide information on their role as conveyor vectors of agents potentially toxic to humans, mechanisms by which they enter the human body, their bioaccumulation, and health human effects.

Keywords: microplastics, nanoplastics, conveyor, gut, microbiota, inflammation, health

1. Introduction

At present, globally, there is an important and relevant environmental and public health problem; 8300 million metric tons of plastic were manufactured worldwide between 1950 and 2017, and this production continues to increase, reaching 390.7 million metric tons only in 2021, and is expected to increase to 34,000 million metric tons by 2050 [1, 2]. Degradation resistance is one of the most critical characteristics that initially gave an advantage to the use of plastics; however, now, it is a significant disadvantage since they are resistant to chemical, biological, and corrosive degradation; therefore, their durability in the environment is greater [3]. Plastics are widely used in various consumer products because of their low density and costs; there are about 30,000 types of plastics, being the most widely used polypropylene (PP), polyethylene (PE), polyethylene terephthalate (PET), polystyrene (PS), polyurethane (PU), polyvinyl chloride (PVC), and polycarbonate (PC) [4]. Plastic residues are transported by rivers, storms, and strong winds or are discarded directly into terrestrial or aquatic ecosystems. Plastic residue pieces are classified according to their size into mesoplastics (50–200 mm), macroplastics (200–1000 mm), and mega-plastics (>1000 mm) [5]. These residues undergo physical changes due to environmental interactions, such as fragmentation, and changes in their physicochemical properties, generating new types of micropollutants such as microplastics (MP (less than 5 mm)) and nanoplastics (NP (<1 μm)) [6]. These tiny plastic particles are ubiquitous worldwide and generate great concern for environmental and human health damage. The impact on human health may be due to their small size, specific surface, and high biological penetrability [7].

Micro(nano)plastics absorbs and transports external toxic pollutants; these plastic particles harm human and wild health, altering the physiological functions of immunity and metabolism and modifying the intestinal microbiota, thereby facilitating exposure to pathogens [8]. In fact, according to morphology, size, and concentration, MP can trigger ecotoxicological problems in different organisms [6]. Exposure to MNP causes local inflammation, oxidative stress, metabolic alteration, gastrointestinal toxicity, hepatotoxicity, reproductive disorders, and neurotoxic effects [9]. Humans are subjected to prolonged exposure to this type of particle in low concentrations; however, these effects on the organism need to be profoundly and widely studied, mainly in the case of NP [9]. So far, eradicating plastic waste remains challenging, and its impact on health is becoming increasingly evident. Despite the increase in recent research on plastic waste, it is still in the early stage; therefore, more research is required. This chapter presents and discusses the role of these particles as carriers of different molecules or microorganisms that directly impact health. Besides, it presents the current knowledge about MNP human exposure pathways and the toxicological effects on the intestinal microbiota-immunity, reproductive and neurotoxic.

2. From plastics to micro- and nanoplastics

The term "plastic" refers to any material with high polymer content as the ingredient; this discovery dates from the beginning of the twentieth century [10, 11]. Plastics are formed by a set of polymers such as polypropylene (PP), polyethylene (PE), polyethylene terephthalate (PET), polystyrene (PS), polyurethane (PU), polyvinyl chloride (PVC), polycarbonate (PC), polymethyl methacrylate (PMMA), polyamide (PA), polylactic (PL) and additives such as stabilizers, flame retardants, plasticizers, and pigments [12].

Plastic presence is ubiquitous in the environment due to its excessive use and polymeric materials' long permanence in plastic residues. When exposed to the environment, plastic undergoes an abiotic degradation, either by a physical or by a chemical weathering process; in the physical weathering process, it suffers from mechanical wear due to water and wind or photodegradation processes, and in the chemical weathering process, hydrolysis or oxidation of polymers takes place [3]. As regards biotic degradation, it is essential to mention that plastic debris in the aquatic ecosystem is usually resistant to corrosion and degradation by microorganisms, being fragmented into smaller-sized plastic particles by the already mentioned abiotic factors [13]. The impressions of the abiotic physical process led to the wear of the plastic parts forming smaller particles, the microplastics (MP) and the nanoplastics (NP). Nowadays, plastic pollution is a global problem that has caused the generation of massive amounts of such MNP, which can be transported through rivers, seas, air, and rain; the last one causes the particles in the air to fall into different areas [14, 15].

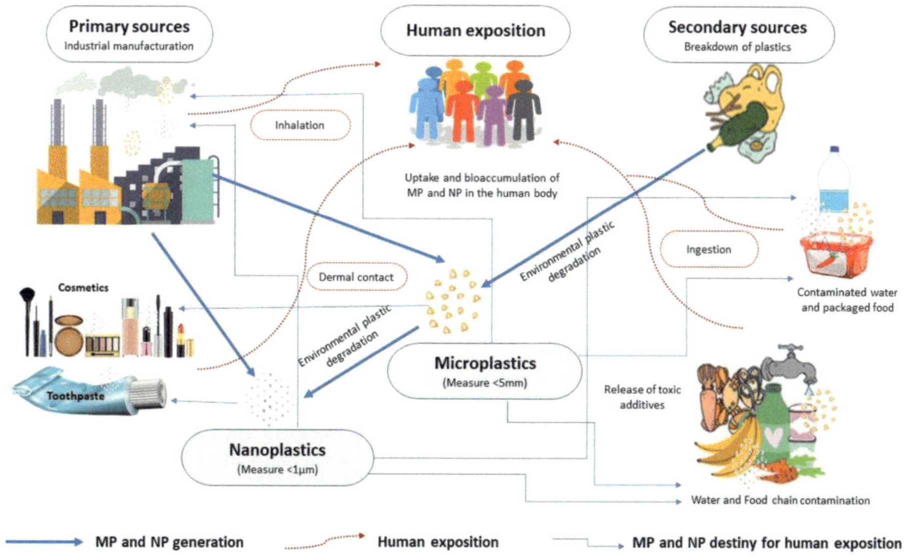

Figure 1.
Microplastics (MP) and nanoplastics (NP) generation and human exposition to these plastic particles.

Based on the source, MNP can be classified as primary and secondary. Primary are particles manufactured for indirect or direct use of raw materials with a size between 0.001 and 5 mm. In contrast, secondary are particles that have been manufactured from macroplastic degradation, defined as plastic material of less than 0.001 mm (<1 μm). Both particles are part of cleaning products, coating, cosmetics, and medical applications; these particles are also generated when bottles, clothing, tires, and containers begin their decomposition by the action of the environment [16–18] (**Figure 1**).

Due to the characteristics of MNP, mainly their size, they become difficult to detect. Indeed, there are still methodological problems with regard to standardization in the characterization and quantification of plastic particles in different materials, media, ecosystems, and the human body [19].

3. Plastic particles as conveyor vectors

The fate of plastic, once discarded, is to follow a long journey. Ultimately, they flow into local, regional, or global ocean currents and usually travel to the ocean gyres, where large amounts of debris accumulate [20, 21]. Indeed, plastic pollution has been found in rivers such as those of the tropical Andes or Germany. Given the above, plastic has the quality of being able to move through great distances and transport different plastic additives used for their manufactures, such as flame retardants, vinyl acetates, styrenes, phthalates, plasticizers, and phenols, which are related to carcinogenic or mutagenic effects as shown in the examples in **Table 1** [27].

Additionally, besides their size, plastic characteristics change because of weathering; this is a change in the microtopography of plastic; for example, smooth areas can become rougher, and cracks, bumps, and bumps cavities are formed, changing to irregular particles with a greater contact surface. These characteristics, together with hydrophobicity, provide a non-polar surface where various pollutants, such as heavy metals, pesticides,

Chemical	Function	Possible effects	Authors
Bisphenol A (BPA)	It is used in polycarbonate plastic and epoxy resin production	Endocrine disrupter	[22]
Ester phthalates (di(2-ethylhexyl)phthalate (DEHP), di-n-butyl phthalate (DBP), and diethyl phthalate (DEP))	Plasticizers Increase plastic flexibility, especially PVC Solvents Fixatives of essences (perfumes and cosmetics)	Some are toxic for reproduction. Others can cause damage in high doses.	[23]
Nonylphenol (NP)	Antioxidant Plasticizer Stabilizer	Highly toxic to marine life. Endocrine disruptor in fish, where it can cause feminization. Toxicity to reproduction and development of other animals.	[24]
Polybrominated diphenyl ether (PBDE)	Flame retardant in plastics, foams, and textiles	Possible endocrine disruptor, especially for thyroid function. Toxic effects on neurodevelopment behavior, immune system, and blood in humans.	[25]
Biphenyls (Polychlorinated biphenyl (PCB))	Flame retardants Plasticizers Insulators	Toxic effects on neurodevelopment, immune system, and reproduction in humans It can produce certain cancer types.	[26]

Table 1.
Main additives in plastics and their effects.

polyaromatic hydrocarbons, antibiotics, fertilizers, and microorganisms, can adhere to and can be harmful to living beings when they are absorbed (**Table 2**) [5, 31, 32].

Indeed, MNP are substrates for colonization by microorganisms and the formation of biofilms [33–36]. These biofilms are communities of microorganisms that adhere to the surface of the plastic and form an extracellular matrix that protects them from adverse environmental conditions, changing the chemical characteristics of the piece, as we will explain widely in the next section. In conclusion, weathering processes change physical and chemical properties of MNP, leading to changes in the environmental behavior of MNP. It is essential to know that these changes in MNP affect the transport and end up in large amounts of toxic contaminants. In addition, weathering affects sedimentation, ingestion by other organisms, and pathogen transportation [15]. Considering this behavior, MNP can be transport vectors for different toxic agents [37].

3.1 Toxic agents carried by plastic particles

Micro(nano)plastics with various functional groups have been considered to interact with organic and inorganic pollutants; size and electrically neutral surfaces allow

Pollutant	Function	Possible effects	Authors
Polycyclic aromatic hydrocarbons (PAHs)	Result of incomplete burning of fossil fuels Ingredients of fuels and tar	All are persistent and bioaccumulated; some are carcinogenic, mutagenic, and toxic to reproduction.	[28]
Pesticide residues	Control various pests and disease carriers Insecticides in agriculture and urban area. DDT use is limited to malaria control.	DDT is highly toxic to aquatic life, a possible endocrine disruptor, and toxic to reproduction. Hexachlorocyclohexane (HCH) is toxic to the liver and kidneys and may act as an endocrine disruptor and a human carcinogen.	[29]
Metals and metalloids	Many uses for the construction industry. Electronics and area of medicine . Machinery, refractory, and automobile industries. Decorative products.	Suppress the immune system, damage the endocrine system, and cause reproductive dysfunction.	[29]
Microorganisms	Play a leading role in countless natural processes, e.g.; they help produce foods, treat wastewater, creating biofuels and a wide range of chemicals and enzymes. Still, they can also cause many essential diseases.	Harmful to health, such as *Escherichia coli* and *Enterococci*.	[30]

Table 2.
Plastic particles transport the main toxic agents.

these particles to be easily transported in aqueous media, which allows MNP to be considered vehicles of importance in the long-range transport of pollutants. In recent years, several research have been developed to evaluate the impact of MNP absorption on human and animal health; being essential to know the distribution and transport mechanism in marine environments, in freshwater, as well as in food chains [38–40]. However, the environmental and biological toxicity of MNP is the genesis of current studies because, to some extent, there are pollutants, which are toxic agents that are adsorbed on the surface of MNP [15, 41].

Rodrigues and coworkers describe that the sorption of a compound can refer to two separate processes: absorption and adsorption. On the one hand, absorption refers to the coupling between a compound and a sorbent mediated by low Van-Der-Waals forces, which allows dissolved molecules to be retained by the sorbent. On the other hand, adsorption needs gamma forces, from Van-Der-Waals forces to ionic or covalent bonds concerning the adsorbent surface [33, 35, 36, 42]. On the one hand, the physical and structural characteristics of MNP are determinants for their transport; however, they also allow the development of absorption and adsorption processes of organic and inorganic compounds on the plastic particle, as well as interfere with the bioavailability of contaminants and their effects on organisms [43, 44]. On the other hand, the characteristics of the contaminants, such as heavy metals, present in plasticizers, stabilizers, pigments, fillers, diluents, solvents, and other additives, as well as in the medium where the MNP particles lie, should also be considered, which may ultimately influence the absorption and adsorption behavior [45, 46].

Furthermore, the characteristics of plastic residues such as size, shape, and polymorphism, as well as environmental conditions (pH of medium, ionic strength, organic matter, microorganisms, and temperature) are involved in MNP pollutant interactions, as well as in hydrophobic and electrostatic interactions [47]. When trying to remove MNP particulate debris by membrane filtration, particle size, biofilm formation, and the interactions described above influence the final fate of these environmental elements [47, 48]. So, MNP undergoing physical, chemical, and biological interactions are released into the environment with the sorption capacity of contaminants that eventually accumulate in the environment in the form of metabolic by-products, degradation by-products, or residual MP [49].

Relevant studies that distinguish between the effects of the synthetic polymer itself and the incorporated additives or chemicals in the polymer itself are still scarce [50]. As we mentioned above, MNP have other constituents (e.g., plasticizers, stabilizers, pigments, fillers, extenders, solvents, and other additives), which have the potential to influence their absorption/adsorption behavior [45, 46, 51]. The following section describes how MNP carries the environment's most frequently encountered toxic agents.

3.2 Microorganisms

The chemical qualities already mentioned of MP, their size, and their physical changes brought about by weathering cause different organisms to adhere to the surface of microplastics. Cracks in the pieces and the hydrophobicity of MP give rise to the adhesion of microorganisms and, consequently, to the formation of glued microorganism biofilms; this ecosystem developed in MP has been named the plastisphere [52]. Plastisphere causes changes in the physicochemical properties of plastic, having a higher level of adsorption pollutants that can be transported using the currents reaching other sites. It should be noted that these biofilms increase the probability of MP ingestion by animals since they attract organisms that depend on chemoreceptors to select prey through olfactory and taste cues [5, 53].

Since the discovery of biofilm-forming microorganisms on MP in marine ecosystems by Carpenter and Smith [54], little research has been done to describe the named plastisphere in 2013. From this date, the research on the microbiota associated with plastic debris has started to receive attention. In the beginning, the marine plastisphere was described, which comprised diverse microbial communities, including opportunistic pathogens transported through MNP worldwide [13, 55].

Marine plastisphere differs from the surrounding water, natural and artificial substrates, and within the plastic types [13, 52]. For example, the microbiota colonizing MP and NP from different polymers in the Bay of Brest (France) presents a prevalence of *Pseudomonadales* and *Oceanospirillales* on PE, alpha-*Proteobacteria* on PP, and *Rhodospirillaceae* on PS [56]. *Pseudomonas* predominate on fresh and marine water microbiota associated with MP and NP; now, n their capability to degrade a wide array of plastics is well known [57]. In addition, *Firmicutes* and *Proteobacteria* were also detected, confirming that some microorganisms may use MP as a raft to migrate from one ecosystem to another [58, 59]. A little research has been done on fungal communities associated with MNP.

The potential role of MNP in dispersing pathogenic species such as *Vibrio*, *Pseudomonas*, *Acinetobacter*, and even the severe acute respiratory syndrome coronavirus 2 (SARS-CoV-2) virus has been reported [58, 60]. For the last case, interaction mechanisms between MNP and the SARS-CoV-2 RNA fragment involve electrostatic and hydrophobic forces, and the interaction affinity is associated with the inherent structural parameters of the MNP. Humans are exposed to SARS-CoV-2-contaminated MNP via their lungs [60].

The transportation feasibility of pollutants and pathogenic microorganisms needs to be better understood by MNP across the waters and their introduction into the food web. Nonetheless, aquatic organisms' ingestion of MNP has been described as a possible transport mechanism of hazardous substances and microbial communities associated with MNP [61].

3.3 Heavy metals (HM)

Not only can microorganisms be transported by MNP, but they can also transport other hydrophobic organic contaminants due to their hydrophobic nature and increased available sites on their contact surface. They can also adsorb heavy metals [62]. Although the mechanisms are being studied, it is observed that the biofilms formed in the MP increase the adsorption of heavy metals through electrostatic interaction, cationic ion exchange, and the formation of complex groups with the functional ones in the biofilms [63, 64].

The first reported interaction between plastics and heavy metals (HM) took place as early as in 2010. Later, some studies proved that HM (Cr, Co, Ni, Cu, Zn, Cd, Pb, Ag, and Hg) could be enriched in MP from different polymers such as PE, PP, PS, PA, PVC, and polyformaldehyde (POM). Interestingly, the adsorption of heavy metals on pristine MP without surface modification is almost negligible, while the eroded/weathering MP, as well as those modified through the attachment of organic matter, accumulate HM [41].

Heavy metals, non-biodegradable inorganic pollutants, cause direct toxicity in the organism; exposure to these pollutants has been linked to the development of cancer and chronic diseases due to their bioaccumulation [65, 66]. As both pollutants (MP and HM) are persistent in environmental degradation and transformation, being resistant and challenging to eliminate, the threat posed by their combined exposure to ecosystems and human beings can be worrisome. A limited number of studies have been conducted on the combined effects of MP and HM, suggesting that the interaction between the two types of pollutants may trigger synergistic, antagonistic, or potentiating effects on organisms. However, since assessment of co-exposure risks in organisms is complex, their effects on humans still need to be determined [41].

3.4 Polycyclic aromatic hydrocarbons (PAHs)

Polycyclic aromatic hydrocarbons are a class of chemicals that occur naturally in coal, crude oil, and gasoline. PAHs have been detected in surface water, drinking water, and wastewater treatment plants, thus representing a risk to public health and the environment [67–71]. Epidemiological studies have demonstrated that exposure to polycyclic aromatic hydrocarbons (PAH) causes damage to human health, ranging from decreased immune response, altered thyroid function, liver, and kidney disease, altered lipid and insulin metabolism, involvement in the development of cancer and non-Hodgkin's lymphoma, and may even negatively influence reproduction and human development. They also have a high carcinogenicity and mutagenicity index, and some cases have been reported where the descendants of mothers exposed to these pollutants have alterations in neuronal development. Therefore, it is crucial to determine the mechanisms of PAH adhesion or absorption in MNP [72, 73].

The interaction of MNP with emerging contaminants of clinical interest and importance for their health effects (per- and polyfluorinated substances (PFAS), PAHs, and BCP) are recalcitrant to environmental degradation. Industrialization development and anthropogenic activities are the primary sources of PAHs in the environment. There are three principal anthropogenic sources of PAH emissions. One is pyrolysis, which includes the incomplete combustion of coal, wood, petroleum, and organic polymer compounds. Another is the leakage and discharge of oil during the mining, transportation, production, and use of crude oil, coal tar, asphalt, shale oil, carbon ink, and industrial mineral oils. The third includes food cooking processes, municipal waste incineration, and agricultural surface emissions (sewage discharge, use of pesticides and herbicides during agricultural irrigation) [2, 74].

Polycyclic aromatic hydrocarbons are transported in the environment in different ways, including leakage, volatilization, and biosynthesis. Like MNP, PAH can also migrate through rivers, precipitation, and air [2, 74]. The main factors affecting the adsorption of PAH by MNP are related to their physical characteristics, including spatial structure, particle size, specific surface area, crystallinity, and glass transition. As regards the chemical interaction between MP and PAH, the major forces are π-π interaction and halogen or hydrogen bonding [75].

Another point of biological and environmental importance to consider in the case of PAHs is their phototoxicity, as the PAH organism exposure in the environment is inevitably accompanied by sunlight. Studies based on various aquatic organisms and mammals have shown that ultraviolet (UV) light amplifies the toxicity of PAH and cannot be neglected [2, 74]. The phototoxicity of PAH in humans has been done *in vitro* at the cellular level, where benzo (*a*) pyrene (BaP) is a hot research topic. Under ultraviolet A (UVA) and ultraviolet B (UVB) irradiation, BaP significantly promoted the production of hydrogen peroxide (H_2O_2) from human epidermoid carcinoma cells (A431) and human primary keratinocytes in a dose-dependent manner [76]. UVA and BaP showed a 7-fold increase in synergy in comparison to UVB. Another study also proved that BaP and its metabolites induced the production of reactive oxygen species (ROS) in human skin keratinocytes (HaCaT) under UVA irradiation, which mediated the formation of intracellular lipid peroxides in a low dose-response manner [77].

3.5 Pesticides

Pesticides have favored the generation of economic income from commercializing agricultural products; however, they are potentially harmful to the environment and human health [78]. These pesticides can bind to organic matter, to PE (polyethylene) agricultural films, or to the soil's clay mineral fraction, making their eradication in the ecosystem difficult. Carbofuran and carbendazim (CBD) are pesticides that are potentially harmful to human health by inhibiting the acetylcholinesterase (AChE) activity, damaging the parasympathetic nervous system, and causing bradycardia, abnormalities in blood pressure, bronchoconstriction, gastrointestinal hypermotility, among other disorders. Unfortunately, these pesticides can adsorb on the surface of MNP, thus becoming internalized in the human organism and increasing the toxicological risk [79]. The pH, ionic strength, and temperature are parameters that are considered to evaluate the interactions between plastic particles and pesticides [80].

It has been considered that the interactions between these compounds with plastic particles start in the farmland and thus are maintained during transport to the ocean. Therefore, it is necessary to consider the search for new strategies to reduce the presence of MNP in the environment since they are vectors that can affect human and animal health and irreparably damage our ecosystems.

3.6 Antibiotics

As mentioned above, in the aquatic environment, MNP are colonized by microorganisms, mainly bacteria that form dynamic biofilms. These biofilms promote antibiotic resistome (AR) in the plastisphere. AR is conformed mainly by antibiotic-resistant bacteria (ARB) and antibiotic or multi-drug-resistant genes. MP have a specific function in the enrichment and transportation of AR through processes such as horizontal gene transfer, conjugation, gene transduction, and transformation. The levels and types of mobile genetic element (MGE), especially integrase genes (*intI1* and *intI2*), besides the structure and density of bacterial communities, the incubation period, and anthropogenic chemicals promoting the co-selection are all significant factors in the development, enrichment, and transportation of plastisphere-associated AR and pathogens in the aquatic environment [81].

Several studies have reported that tetracyclines, macrolides, fluoroquinolones, sulfonamides, aminoglycosides, chloramphenicol, and beta-lactams are frequently encountered antibiotics adhered to MP detected in worldwide marine environments [82]. AR associated with MP (AR-MP) can be a vector for bioaccumulation of these antibiotics and have higher lethality than MP [83].

Exposure to AR-MP and pathogens could be linked to disruption in food chain, enzymatic and genetic toxicity, oxidative stress, altered feeding behavior, gut microbiota dysbiosis, metabolic disorders, pathogen-induced diseases, and drug resistance [81].

The adsorption of antibiotics on microplastics can span a very wide transport range and cause a large combination of effects. One study investigated the adsorption of mainly five types of antibiotics (sulfadiazine (SDZ), amoxicillin (AMX), tetracycline (TC), ciprofloxacin (CIP), and trimethoprim (TMP)) on five types of microplastics (polyethylene (PE), polystyrene (PS), polypropylene (PP), polyamide (PA), and polyvinyl chloride (PVC)) in freshwater and seawater systems. It was shown that PA had higher adsorption capacity for antibiotics in the freshwater system, which can be attributed to its porous structure and the ability to form many hydrogen bonds [83].

In resume, from the time we used plastic products, they have had a journey through tributaries, suffered weathering in the sea, they journeyed through the chain trophic and the air, which causes them to absorb contaminants on the way, becoming a conveyor vector.

4. Human exposure to plastic particles

Plastic particles, as mentioned above, have become a multifactorial and far-reaching problem; this does focus not only on the nutritional consequences or damage to aquatic systems but also on the epidemiological consequences [84, 85]. Microplastics and nanoplastics (MNP) are ubiquitous in the environment, and the effects of human exposure to these particles have not been fully described, representing a challenge and point of studies related to human health [86]. Even exposure to low concentrations

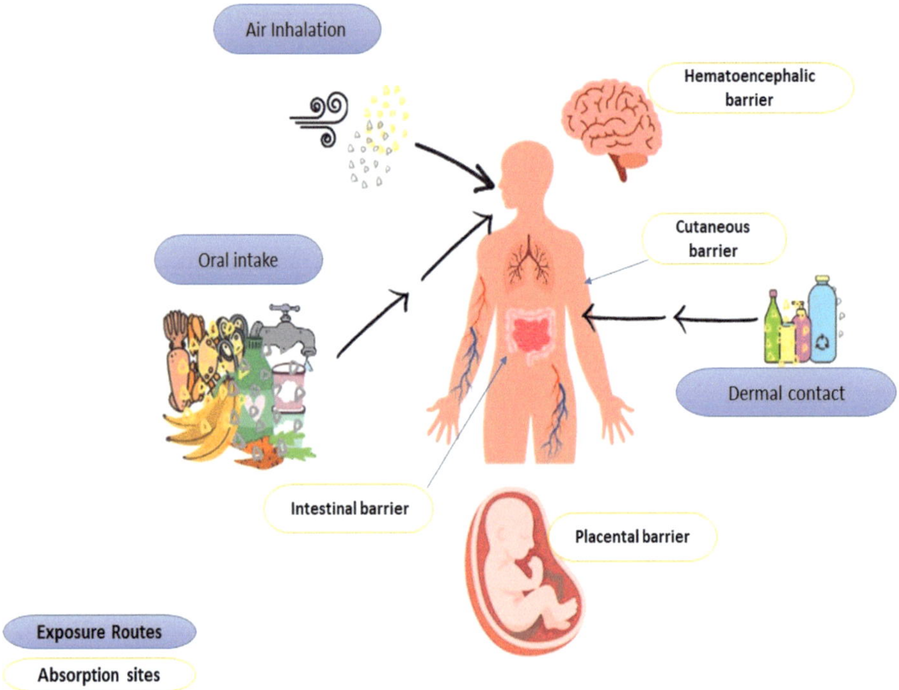

Figure 2.
Human exposure routes to plastic particles and sites where their absorption occurs.

of these particles for prolonged periods is a worrying situation, as they are present in food and drinking water [87].

The Center for Diseases Control and Prevention (CDC), in [88], verified that the main routes of exposure to MNP for humans are inhalation, ingestion, and dermal contact (**Figure 2**). According to data reported by Cai and coworkers in [89], marine animals ingest MNP contained in water, and as they are eaten, they can be passed mainly in this way to humans; likewise, MNP have been found in table salt [90]. Moreover, other MNP exposition vias also occur by ingestion of food and plastic packaged consumer goods, which can become contaminated by package degradation or by ingestion of contaminated food or water and inhaling contaminated air; for all vias, it has been determined that they impact negatively human health [91] (**Figure 2**).

The main absorption sites are related to the principal body barriers (**Figure 2**). Contact exposure begins with penetration through the pores of the skin, ingestion through the consumption of MNP-contaminated water or food, articles of personal use, etc., which pass the intestinal barrier. Exposure via area or inhalation is brought about by the presence of MNP in the air, and through breathing, enters the body and accumulates in the respiratory system. The three main routes of exposure to MNP are described in more detail below.

4.1 Exposure by ingestion

The most evaluated and common route by which humans are continuously exposed to MNP is the oral route; it occurs through the intake of contaminated water, food, and personal cleaning items such as toothpaste. Extensive use of plastic components in drinking water networks and filtration of particles from water pipes under long-term use can lead to high exposure to large concentrations of MNP that can impact health [92]. Several studies have proven that humans consume hundreds of millions of MNP particles from packaging tea bags, seafood, table salt, honey, beer, and bottled beverages. In food, MNP can be present in animals that become contaminated through their environment or food chains, such as shellfish. MNP absorbed by microorganisms lower in the food chain are ingested by higher organisms in it and are finally ingested by humans affecting human health through bioaccumulation [93, 94]. MNP can also be present in contaminated food during its production or packaging processes [95–98].

Deng et al. [99] used pristine particles of fluorescent polystyrene microplastics with two diameters, 5 and 20 μm, to evaluate their distribution in the tissues, their accumulation, and the specific effect on the health of the mice that formed both experimental groups. The results indicated an accumulation of particles in the liver, kidneys, and intestine; the distribution of these particles depends on the size. It was determined that the intake of MP alters energy and lipid metabolism and intervenes in oxidative stress. It has also been determined that particles up to 150 μm can move through the mammalian intestinal barrier and that the absorption rate is <0.3% [100]. On the other hand, Campanale et al. [101] report that most particles down to 10 μm can penetrate all organs, including the brain; however, the consequences have not been established.

Due to the permanence and accumulation of MNP in the food chain and in wildlife, whose resources are destined for human consumption, they are of great relevance because, as we depict above, they act as transporters of pathogens, pollutants, and potential toxins with negative impacts on health [102]. In addition, the accumulation of plastic particles at all trophic levels potentially exposes humans [103]. The European Food Safety Authority (EFSA) and World Health Organization (WHO) suggest that exposure to MNP in humans has low adverse effects; however, more evidence is needed on the preliminary signs, which propose better analytical methods to assess this problem. Auta and collaborators [104] studied the toxic effects of exposure to MNP in marine organisms, observing that depending on the type of organism, there is bioaccumulation of particles, metabolic changes, inflammatory processes, reproductive problems, changes in behavior between others, and all of this will determine the ability to survive in nature.

The lack of an established and validated method providing non-destructive evidence of the presence of MNP in the tissue limits estimates of the degree and effect of their exposure [103]. Ultra-thin tissue fractions are used in medical research, making assessment of the possible presence of MNP and their influence on disease processes even more challenging. It is essential to deepen studies on the possible immunological effects, gastrointestinal alterations derived from modifying the intestinal microbiota due to exposure to these particles, and the damage they generate. Murine models are a tool of choice to evaluate these processes, which allow us to know the risk to health in animals and humans and evaluate the sensitivity of specific pathologies to such exposure.

4.2 Inhalation exposure

Inhalation itself is considered one of the main human exposures to MNP; it also occurs because these particles are found in the air, some of the sources of which can be synthetic textiles, tire erosion, and some types of polymeric fibers such as PP, PE, PS, and PET. The exposure per individual has been estimated at 26,130 airborne plastic particles daily [105].

Once inhaled, MNPs reach the respiratory epithelium, causing their translocation by diffusion, penetrating the cell directly or by uptake of MNPs by endocytosis or phagocytosis. In the alveoli, phagocytosis is carried out for particles between 1 and 3 μm in size, being this mechanism the main one for this route of exposure. Smaller particles could be passively transported by diffusion across the membrane [106]. However, the number of studies evaluating MNP in indoor and external environments is scarce [107].

After studying MNP presence in the air of private residences and public offices, it was reported that indoor air contains a higher concentration of MNP than outdoor air, implying that exposure in humans is higher [108]. Likewise, MNP can be inhaled through fine dust suspended in the air and then release chemical additives [109]. Plastic fragments and fibers are the most common means that favor the inhalation of MNP [46].

Removal of inhaled particles can be done through mucociliary transport, resulting in minimal MNP concentration in the airways or phagocytosis by macrophages. However, it has been reported that MNP can circumvent these mechanisms by accumulating in the lungs and entering the systemic circulation [110]. Another conduit through which inhaled nanoparticles can reach the central nervous system (CNS) is the olfactory bulb [111]. Some studies indicate that in some neurological diseases, the absorption of plastics in neonates (plasticenta) and its consequences are associated with inhalation to MPN [112]. One of the studies by Fournier and collaborators [110] shows that an intratracheal dose of microplastics during gestation undergoes maternal-fetal translocation, generating secondary damage to fetal development.

4.3 Dermal exposure

Topical exposure to MNP contained in personal hygiene products and contaminated water affects skin health, and epidermal cells have shown stress in oxidative experiments *in vitro* [113]. Tiny synthetic fibers (<25 μm) penetrate skin pores measuring around 40–80 μm and will bypass the stratum [114, 115]. However, nanoparticles smaller than 100 nm can be exclusively absorbed in the stratum corneum [100]. Exposure by this means is based on individual susceptibility because the pores of human skin have different characteristics between one living being and another.

The nanotoxicology of nanoengineered materials indicates that nanoparticles (NPs) with a size <40 nm can enter the body through the epidermal barrier [116, 117]. This exposure occurs when there is contact between the dermis and NP in the environment, which can occur when showering with water or using personal care products containing, for example, nanopolystyrene [31, 118]. Indeed, the Food and Drug Administration (FDA) passed the microsphere-free water law that went into effect for products manufactured after the 2017–2018 law that restricts the use of MP in exfoliating products, toothpaste, and other cosmetics to reduce dermal exposure and decrease residues in water systems. In the medical field, there is also a wide use of MP for drug administration; it is considered that people may have been exposed to 4594–94,500 MP particles in just 5 mL of the product [35].

Considering the seriousness of plastic pollution in recent decades, it is necessary to act and raise awareness about its effects on the human body. More research is currently needed to analyze whether there is a risk of absorption of MNP through the dermal exposure route in humans. For example, it has not been determined whether rigid MNP cross the subcutaneous barrier under normal conditions; however, it has been shown that they accumulate in hair follicles, and it was observed that Langerhans cells absorbed them [119]. The following section describes the most critical effects of MNP exposure on human health.

5. Human digestion of toxic agent-contaminated MP and NP

When we eat food with MNP contaminants, they can be released into the digestive system. The contaminants can then be absorbed through the lining of the gastrointestinal tract (GIT) and transported through the circulatory system to various organs in the body [120]. The exact process has yet to be fully understood, but it is believed that contaminants are released by the action of stomach acids and digestive enzymes, as well as by interaction with gut microbiota. So far, research on MP digestion is scarce; it has been performed *in vitro* and using single-batch models, and studies of MP with adsorbed toxicants are even scarcer.

An example is the study on MP with Chromium (Cr) in a human static *in vitro* digestion model, including mouth, gastric, small intestinal, and large intestinal digestive phases where no microbiota was included. The results showed high desorption under acidic conditions, which means that Cr adsorbed on MP with different polymers, such as PE, PP, PVC, PS, and PLA, 150 μm, was more bioaccessible due to acidic conditions favoring the desorption of anionic Cr species from the MP surface. At the same time, no release was observed in the oral phase. Additionally, PLA exhibited the highest oral bioaccessibility compared to other polymer types, probably due to its degradation enhanced by the action of enzymes in simulated digestive juices [121].

Moreover, it has been reported that there is no striking alteration in the physicochemical characteristics of the five types of MP (PE, PP, PVC, PET, and PS) by artificial digestive juices mimicking the saliva, gastric, and intestinal phases of human digestion. However, corona formation on the MP surface due to the adsorption of organic compounds, such as proteins, mucins, and lipids during digestion, should be considered [122].

Increasing attention is paid to the gastrointestinal tract (GIT) as the first barrier and a portal of entry and target for MP. As we have reviewed above, MP harbors a unique microbiome shaped by polymer type and environmental factors. After MP ingestion, several human pathogenic microbial species associated with MPs represent a potential risk for humans. In addition, ingestion of AR-MP could enhance the antibiotic release and the development of antibiotic-resistant bacteria in the human gut acting as vehicles for transferring antimicrobial resistance genes to the intestinal microbiome and favoring human infection related to resistant bacteria. However, reliable *in vitro* gut models could then be made relevant to mimic such exposure scenarios considering the consequences on at-risk populations (children and elderly) and, ultimately, pathophysiological conditions (obese patients and inflammatory bowel disease patients) [123].

6. Principal health effects associated with plastic particles' exposure

As mentioned before, there is still a lack of definitive evidence linking the consumption of MNP with human health; in correlation studies on people exposed to high concentrations of microplastics, experiments with animal models and cell cultures have shown that microplastics can cause the activation of the immune response, stress and induce increased or developed toxicity.

6.1 Effects on physical barriers

Ingestion is considered the main route of exposure to MNP, so the tissues of the human gastrointestinal tract are considered the most exposed, data supported by changes in the intestinal microbiota. Information has been found on morphological alterations in Peyer's plaques, which are part of the lymphoid tissues associated with the intestine, leading to the activation of inflammatory processes. Other cells that undergo morphological and functional changes are the M cells responsible for the absorption of particles from the intestinal lumen to the basolateral region, where many lymphocytes and others are found. Cells of the immune system; These, in turn, transport antigens in an integral way because on their surface, there are receptors for specific antigens, proteins, viruses, and bacteria incorporated into endocytic vesicles in the luminal plasma membrane and transported to pre-lysosomes and lysosomes; site of presentation of antigens and their epitopes to dendritic cells, macrophages, and lymphocytes present in the epithelium of lymphoid follicles or in the mucosa; this feature of a carrier cell also allows it to transport plastic nanoparticles from the intestine to lymphoid tissues; therefore, it is associated with pro-inflammatory processes in humans [31, 101, 124]. These particles can sometimes concentrate and disrupt immunity processes in the intestinal barrier.

The persorption process for larger particles up to 130 μm in diameter, performed by the epithelial cells of the gastrointestinal tract, is another pathway of uptake but not absorption of MNP.

In addition to the particle size of MNP, their chemical composition, shape, and hydrophobicity are other factors that affect absorption and translocation to other organs. However, it has been determined that hydrophobic surfaces can be transported more efficiently through the intestinal mucosa. These characteristics allow corona formation by the adhesion of microorganisms, proteins, and other biomolecules [125]. MNP, after being internalized in the intestinal epithelium, can be released into the intestinal lumen due to tissue cell renewal that occurs approximately every other day; therefore, it has been determined that they do not reach the bloodstream. Some studies assume that the continuous accumulation of particles in the liver and digestive tract generates toxic effects [126]. Li et al. [127], in a mouse model, fed PE particles, reported increased inflammation in the small intestine, followed by changes in the microbiota and an increase in systemic pro-inflammatory markers.

6.2 Effects on gut microbiota (GM)

The gut microbiota plays a vital role in human health; this term refers to the microbial ecosystem that colonizes the gastrointestinal tract. The relevance and impact of resident bacteria on host physiology and pathology are documented, as are the main functions of GM ranging from metabolic activities to their translation into metabolic activities, in the recovery of energy and nutrients, protection of the host against foreign microorganisms; trophic functions essential for the development and homeostasis of the immune system; absorption of vitamins, etc. On the other hand, evidence implicates the intestinal microbiota in pathological processes such as multiorgan failure, colon cancer, inflammatory bowel disease, obesity, neurocognitive disorders, and even personality disorders. Variations in GM after *in vivo* exposure to microplastics are being investigated in several contexts. Metagenomic studies have been developed to explore changes in GM; for example, in the springtail *Folsomia candida*, exposed to PVC microspheres of 80 to 250 μm, 1 μg/kg of dry soil for 56 days, intestinal microbial diversity was significantly higher, and its composition showed a significant decrease in *Bacteroidetes* and an increase in *Firmicutes* [128]. On the other hand, in the *crab Eriocheir sinensis*, during 21 days of exposure to PS microspheres of 5 μm at doses of 40 mg/L, a decrease in the relative abundance of *Firmicutes* and *Bacteroidetes* was reported, and an increase in the relative abundance of *Fusobacteria* and *Proteobacteria* [129].

Within the studies carried out in murine models, a large number of significant modifications have been found in the composition of the bacterial phyla of the GM after chronic exposure to PS microspheres of 5 μm at a concentration of 100 μg/ day for 5–6 weeks, the relative abundance of the phylum α-*Proteobacteria* decreased, the relative abundance of the phyla *Actinobacteria* and *Firmicutes* was also reduced [38, 130, 131]. In the study by Luo et al. [132], they used a protocol with similar exposure, used PS beads of 5 μm and a dose of 1000 μg/L for 6 weeks, exposure was performed in pregnant and lactating mice, some transcriptome studies and 16 s RNA sequencing indicated that MP caused metabolic disorder in pregnant mice associated with dysbiosis and intestinal barrier dysfunction. Simultaneously, maternal exposure to MP triggered intergenerational effects and even caused long-term metabolic consequences in generations 1 and 2. MP and PN influence the development of intestinal dysbiosis; these variations can, in turn, cause functional alterations in the immune system.

6.3 Effects on immune response

Plastic particle accumulation has been demonstrated in the liver, kidneys, brain, spleen, and reproductive organs (**Figure 2**) [40, 113, 133]. The intestinal immune system interacts with non-pathogenic commensal microorganisms and harmless food antigens that generally must be immunologically tolerated. In turn, the immune system must retain the ability to respond quickly to infectious toxins. This event depends on the different mechanisms involving myeloid, lymphoid, innate, and T cells found in the intestinal lamina propria and mesenteric lymph node. The immunotoxicity of MNP has yet to be thoroughly evaluated; however, until now, it is known that the immune system is compromised by exposure to these.

There are several examples of immune response after MP exposure. One of this is the exposure in *Daphnia magna* to carboxylate-modified PS-NP with a size of 500 nm at a concentration of 85 mg/L for 1 year, which increases the number of hemocytes. Studies on invertebrates have determined that exposure to PS alters the immune system [134]. After exposure to amino-modified nanoplastics with a particle size of 50 nm at 10 μg/L after 24 h, hemocytes presented mitochondrial and lysosomal alterations; after 72 h, the levels of bactericidal activity and transcription of genes related to the immune system were elevated, and after 96 h of exposure, hemolymphatic phagocytosis, oxidative stress levels, and microbiota were modified [135]. On the other hand, a specific dysregulation of proteins involved in the immune response in the hemolymph of mussels exposed to PE and PA-MP at doses of 845 and 1296 particles/L, respectively, for 52 days was observed. Finally, exposing mollusks of *Mytilus* spp. to a mixture of PE and PS microspheres with a size <400 μm, 10 μg/L, for 10 days, improves the activity of acid phosphatase in hemolymph and causes DNA damage, as published by Revel et al. [136].

After *in vivo* evaluation of the effects of exposure to PS-NP (41 nm, 0.025–0.2 μg/μL) in fish, *Pimephales promelas* modify their neutrophil function. This activity is dose-dependently administered, and myeloperoxidase activity is also modified [137].

As mentioned above, there are intergenerational effects of exposure to MNP, for example, after a 90-day administration of 2 mg/day/mouse of PE particles. In mice, exposure to PE-MP for 5 weeks at administered doses of 20 and 200 μg/g modifies serum interleukin 1α (IL1α) and granulocyte colony-stimulating factor (G-CSF), decreases regulatory T-cell count, and increases the proportion of Th17 cells in splenocytes [83]. In a mouse model of pregnant and lactating female, it has also been reported that blood neutrophil counts and immunoglobulin A (IgA) levels were elevated in mothers and spleen lymphocytes in mothers and offspring [138].

The presence of macroplastics in the gastrointestinal tract of the shark *Scyliorhinus canicula* was associated with a significant upregulation in the expression of T-cell receptor beta (TCRß) and T-cell receptor delta (TCRδ) and immunoglobulin M (IGM) cell receptors in the spleen [139].

Secretory IgA is one of the main components of the immune barrier present on the surface of the intestinal mucosa, so it interacts with symbiotic bacteria to protect against pathogens. On the other hand, under the intestinal epithelium are located different cells of the immune system such as T, B, dendritic cells, and macrophages, which coordinate and are responsible for the presentation of antigen-producing antibodies and secretors of cytokines in the intestinal barrier [140]. Within the research that have been carried out to evaluate the immunotoxicity caused by MNP, it is estimated that in immune cells, they are the target and where toxic effects are present [139].

As the production of plastics increases, so does pollution in the air, water, and food chain, increasing their permanence and exposure in living beings. It is estimated that every 10 g of human feces contains at least 20 particles of MP or NP [141]. In 1 g of tissue obtained from adult colectomy samples, 28 ± 15 particles were determined, and approximately 331 MP particles were sampled [142]. Confirming that MP can enter the gastrointestinal tract and the intestinal permeability caused by MNP allows commensal and pathogenic bacteria to penetrate, causing pro-inflammatory immune responses [143]. The transcription of genes related to cell proliferation, metabolism, and immunity in epithelial cells is regulated by GM [144], so it is necessary to monitor and determine damage to the intestinal barrier as the impact of microplastics on GM cannot be ignored. In addition to the intestinal barrier, studying the effects on the intestinal-vascular barrier (IVB) is relevant, an additional cellular barrier located under epithelial cells, which controls access to circulation and the liver [145]. Suppose any molecule or microorganism crosses the epithelial barrier. In that case, they will remain in the lamina propria, and when the IVB is damaged or exposed, intestinal pathogens can enter the systemic circulation and cause further damage [146]. The damage caused in IVB is brought about by bacterial translocation and MP and NP, thus participating in the pathogenesis of non-alcoholic steatohepatatis (NASH) [147]. However, there is still not enough evidence on the effects of MP and NP on the intestinal-vascular barrier, remembering that the intestinal system is a complex and multifunctional organ. The damage exerted by these particles is not limited to the intestines; they are also causes of inflammation and metabolic dysfunction through intestinal injuries. It remains to be clarified whether acute, long-term toxic effects are triggered in humans.

6.4 Effects on the respiratory tract

Plastic particles are widely distributed in the atmosphere and can be inhaled directly and continuously, posing a potential risk to the respiratory system [148]. The translocation vias of MNP into body compartments have been previously described. Most research to determine the effects of MNP on the respiratory system have been conducted *in vitro* to determine lung cell viability, finding that exposure to these particles and the pollutants they can carry increases cell death, induces oxidative stress and inflammatory response, and increases epithelial barrier destruction. Specific NPs have been reported to reduce the repair capacity of the lungs, leading to tissue damage and lung disease after prolonged exposures [149].

Lu and coworkers [130] administered microplastic droplets (1–5 μm, 300 μ/20 μl) into the nasal passage of mice, where they reported the presence of MP in the airways, alveoli, and interstitial, indicating that particulate matter (PM) can cross the alveolar epithelial barrier. In other studies with animal models, pathological examination showed alveolar structural alteration and disorganized arrangement of the bronchial epithelium because of tracheal administration of PM at a concentration of 2 mg/200 μl [150]. The translocation of MPs and NPs to the pulmonary level activates the immune response by releasing secretory signals (cytokines) because it is estimated that the airway epithelium detects them in the same way as if they were allergens or other irritants. The cytokines whose expression is increased are tumor necrosis factor α (TNF-α), interleukin 6 (IL-6), and interleukin 1β (IL-1β), and transforming growth factor beta (TGF-β) [149–151]. These underlying mechanisms are related to refractory asthma; the aggregate of these data suggests that PMs can differentiate alveolar cells, activate fibroblasts and extracellular matrix organization, events related to fibrosis and pulmonary emphysema and even to lung cancer. Therefore, it is necessary to show the potential risk posed by inhaled MNP.

6.5 Effects on the liver and lipid metabolism

As mentioned above, when responding to oral exposure to MP and NP, multiple groups showed that the gut microbiota is altered, as well as serum and hepatic markers of amino acid and lipid synthesis and metabolism, as well as liver inflammation [38, 83, 127]. Exposure to these particles produces hepatocellular edema and infiltration of inflammatory cells with increased production of IL-1β and TNF-α; hepatotoxic injury also occurs, whose elevation of serum markers of liver function depends on concentration, exposure time, and particle size; mice exposed to 250 nm Polyurethane (PUR) particles for 10 days showed elevated serum levels of alanine transferase (ALT), alkaline phosphatase (ALP), IL-6, and TNF-α, hepatic vascular congestion, and hepatocyte vacuolization [152]. Subsequently, the exposure was brought about with a mixture of PS particles of 1.4 and 10 μm by the oral probe in triple transgenic reporter mice (HOTT) with heme oxygenase-1 (HO-1), which expressed a LacZ indicator sensitive to oxidative stress and pro-inflammatory processes.

Effects that depend on the concentration of particles [122]. The liver is the main site where lipid metabolism occurs; it is sensitive to pathologies such as NASH, which manifests as an accumulation of fatty vesicles combined with cholesterol and high-circulation triglycerides. The MP and NP alter these performances in rodents. Hepatic distension increased hepatic triglycerides, total cholesterol, and decreased expression of peroxisome proliferator-activated receptor alpha (PPARα) and peroxisome proliferator-activated receptor gamma (PPARɣ) has been observed. The lipid-sensitive nuclear receptor PPARα regulates catabolism and fatty acid elimination, and some studies concluded that it may have anti-inflammatory effects by suppressing the nuclear factor-kappa B (NF-κB) [153, 154]. Thus, the regulation of PPARα is reduced, predicting the potential development of NASH. It is understood that changes in lipid metabolism depend on particle size, as mentioned above.

7. Conclusion

When evaluating the effects of MNP in humans, routes of exposure such as absorption, ingestion, and inhalation have been established; however, a few research have described the gastrointestinal and pulmonary toxicity that develops oxidative stress, inflammatory response, and metabolic disorders.

In addition, it is essential to understand whether MNP can be further degraded after ingestion under acidic conditions of the stomach or within cell lysosomes—the long-term occurrence and fate of ingested MP and NP in the human body warrant further investigation.

Given the wide variety of particle sizes, shapes, and chemical compositions of plastics, the potentially dangerous effects of different types of MNP on human health remain largely unknown. Therefore, we recommend that future research focus on understanding the potential hazards and risks of chronic exposure to various types of MNP at relevant concentrations.

Both MP and NP can significantly affect marine organisms and human health. Since MP are small and lightweight, human-consumed marine species can ingest them quickly, accumulating this debris in tissues, circulatory systems, and the brain. Research on MP bioaccumulation and biomagnification through the food web is scarce; efforts should be directed toward clarification.

Compared to MP, NPs can spread even more efficiently in animal bodies and translocate between various organs. They also offer more effective adsorbents, transport of contaminants, and proven threats to human and animal health. However, NPs have yet to be noticed in most studies due to their enigmatic existence, limits on sampling and analysis protocols, and non-standardized basic parameter units, such as particulate matter abundance.

Acknowledgements

This work was supported by Tecnológico Nacional de México (TNM 5669.19-P), CONAHCyT PRONACES (318956) and (317526). Morales-Cano Karla Lizzeth (Ph.D.) and Hermida Castellano Leandro (M.Sc) received a CONAHCyT scholarship (719165 and 1090175).

Conflict of interest

No conflicts to declare.

Notes/thanks/other declarations

The manuscript has been viewed and approved by all authors. All recognized persons have seen and approved mentioning their names in the article. The data in the current manuscript have not been previously published in article form, nor are they in current presentation with any other journal.

Author details

Karla Lizzeth Morales-Cano[1], Leandro Hermida-Castellanos[1], Carlos M. Adame-Adame[2], Luis Alberto Peralta Peláez[1] and Carolina Peña-Montes[1]*

1 Tecnológico Nacional de México campus Veracruz, Food Research and Development Unit (UNIDA), Veracruz, México

2 Hospital Medimac, Boca del Río, Veracruz, México

*Address all correspondence to: carolina.pm@veracruz.tecnm.mx

References

[1] Statista Annual Production of Plastics Worldwide from 1950 to 2020. 2021. Available from: https://www.statista.com/statistics/282732/global-production-of-plastics-since-1950

[2] Sun K, Song Y, He F, Jing M, Tang J, Liu R. A review of human and animals exposure to polycyclic aromatic hydrocarbons: Health risk and adverse effects, photo-induced toxicity and regulating effect of microplastics. Science of the Total Environment. 2021b;**773**:145403. DOI: 10.1016/j.scitotenv.2021.145403

[3] Chamas A, Moon H, Zheng J, Qiu Y, Tabassum T, Jang JH, et al. Degradation rates of plastics in the environment. ACS Sustainable Chemistry & Engineering. 2020;**8**(9):3494-3511. DOI: 10.1021/acssuschemeng.9b06635

[4] Lin L, Zuo L-Z, Peng J-P, Cai L-Q, Fok L, Yan Y, et al. Occurrence and distribution of microplastics in an urban river: A case study in the Pearl River along Guangzhou City, China. The Science of the Total Environment. 2018;**644**:375-381. DOI: 10.1016/j.scitotenv.2018.06.327

[5] Carbery M, O'Connor W, Palanisami T. Trophic transfer of microplastics and mixed contaminants in the marine food web and implications for human health. Environment International. 2018;**115**:400-409. DOI: 10.1016/j.envint.2018.03.007

[6] Xiang Y, Jiang L, Zhou Y, Luo Z, Zhi D, Yang J, et al. Microplastics and environmental pollutants: Key interaction and toxicology in aquatic and soil environments. Journal of Hazardous Materials. 2022;**422**:126843. DOI: 10.1016/j.jhazmat.2021.126843

[7] Liu M, Liu J, Xiong F, Xu K, Pu Y, Huang J, et al. Research advances of microplastics and potential health risks of microplastics on terrestrial higher mammals: a bibliometric analysis and literature review. Environmental Geochemistry and Health. 2023;**45**(6):2803-2838. DOI: 10.1007/s10653-022-01458-8

[8] Kinigopoulou V, Pashalidis I, Kalderis D, Anastopoulos I. Microplastics as carriers of inorganic and organic contaminants in the environment: A review of recent progress. Journal of Molecular Liquids. 2022;**350**:118580. DOI: 10.1016/j.molliq.2022.118580

[9] Sangkham S, Faikhaw O, Munkong N, Sakunkoo P, Arunlertaree C, Chavali M, et al. A review on microplastics and nanoplastics in the environment: Their occurrence, exposure routes, toxic studies, and potential effects on human health. Marine Pollution Bulletin. 2022;**181**:113832. DOI: 10.1016/j.marpolbul.2022.113832

[10] Andrady AL, Neal MA. Applications and societal benefits of plastics. Philosophical Transactions of the Royal Society of London. Series B, Biological Sciences. 2009;**364**(1977-84):23. DOI: 10.1098/rstb.2008.0304

[11] Rochman CM, Brookson C, Biker J, Djuric N, Earn A, Bucci K, et al. Rethinking microplastics as a diverse contaminant suite. Environmental Toxicology and Chemistry. 2019;**38**:703-711. DOI: 10.1002/etc.4371

[12] Kannan K, Vimalkumar K. A Review of Human Exposure to Microplastics and Insights into Microplastics as Obesogens. Frontiers in Endocrinology (Lausanne). 2021;**12**:724989

[13] Barros J, Seena S. Plastisphere in freshwaters: An emerging

concern. Environmental Pollution. 2021;**290**:118123. DOI: 10.1016/j.envpol.2021.118123

[14] Leslie HA, van der Meulen MD, Kleissen FM, Vethaak AD. Microplastic Litter in the Dutch Marine Environment. Providing facts and analysis for Dutch policymakers concerned with marine microplastic litter. 2011. Deltares, report 1203772-000

[15] Sun Y, Yuan J, Zhou T, Zhao Y, Yu F, Ma J. Laboratory simulation of microplastics weathering and its adsorption behaviors in an aqueous environment: A systematic review. Environmental Pollution. 2020a;**265**:114864. DOI: 10.1016/j.envpol.2020.114864

[16] [AMEC] AMEC Foster Wheeler Environment & Infrastructure UK Limited. Intentionally added microplastics in products. European Commission (DG Environment) Doc. Ref 39168 Final Report 17271i3. Oct. 2017

[17] Da Costa JP, Santos PS, Duarte AS, Rocha-Santos T. (Nano)plastics in the environment - sources, fates and effects. The Science of the Total Environment. 2016;**566-567**:15-26. DOI: 10.1016/j.scitotenv.2016.05.041

[18] Hanvey JS, Lewis PJ, Lavers JL, Crosbie ND, Pozo K, Clarke BO. A review of analytical techniques for quantifying microplastics in sediments. Analytical Methods. 2017;**9**(9):1369-1383. DOI: 10.1039/C6AY02707E

[19] Schwaferts C, Niessner R, Elsner M, Ivleva NP. Methods for the analysis of submicrometer- and nanoplastic particles in the environment. TrAC Trends in Analytical Chemistry. 2019;**112**:52-65. DOI: 10.1016/j.trac.2018.12.014

[20] Donoso JM, Ríos-Touma B. Microplastics in tropical Andean rivers: A perspective from a highly populated Ecuadorian basin without wastewater treatment. Heliyon. 2020;**6**(7):e04302. DOI: 10.1016/j.heliyon.2020.e04302

[21] Klein S, Worch E, Knepper TP. Occurrence and Spatial Distribution of Microplastics in River Shore Sediments of the Rhine-Main Area in Germany. Environmental Science and Technology. 2015;**49**(10):6070-6076. DOI: 10.1021/acs.est.5b00492

[22] Tang Y, Zhou W, Sun S, Du X, Han Y, Shi W, et al. Immunotoxicity and neurotoxicity of bisphenol a and microplastics alone or in combination with a bivalve species, *Tegillarca granosa*. Environmental Pollution. 2020;**265**:115115. DOI: 10.1016/j.scitotenv.2021.147003

[23] Meeker JD, Sathyanarayana S, Swan SH. Phthalates and other additives in plastics: human exposure and associated health outcomes. Philosophical Transactions of the Royal Society B: Biological Sciences. 2009;**27;364**(1526):2097-2113. DOI: 10.1098/rstb.2008.0268

[24] Soares A, Guieysse B, Jefferson B, Cartmell E, Lester JN. Nonylphenol in the environment: A critical review on occurrence, fate, toxicity and treatment in wastewaters. Environment International. 2008;**34**(7):1033-1049. DOI: 10.1016/j.envint.2008.01.004

[25] Darnerud PO, Eriksen GS, Jóhannesson T, Larsen PB, Viluksela M. Polybrominated diphenyl ethers: Occurence, dietary exposure, and toxicology. Environment Health Perspective. 2021;**109**:49-68. DOI: 10.1289/ehp.01109s149

[26] Ulbrich B, Stahlmann R. Developmental toxicity of polychlorinated biphenyls (PCBs): A

systematic review of experimental data. Archives of Toxicology. 2004;**78**(5):252-268. DOI: 10.1007/s00204-004-0583-y

[27] Greenpeace Mexico. Microplastic in shellfish and fish. 2016. Available from: http://archivo-es.greenpeace.org/espana/es/Informes-2016/Agosto/Plasticos-en-el-pescado-y-el-marisco/

[28] Cruz-Salas AA, Velasco-Pérez M, Mendoza-Muñoz N, Vázquez-MorillasA,Beltrán-VillavicencioM, Alvarez-Zeferino JC, et al. Sorption of Total Petroleum Hydrocarbons in Microplastics. Polymers. 2023;**15**(9):2050. DOI: 10.3390/polym15092050

[29] Borges-Ramírez MM, Mendoza-Franco EF, Escalona-Segura G, Osten JR. Plastic density as a key factor in the presence of microplastic in the gastrointestinal tract of commercial fishes from Campeche Bay, Mexico. Environmental Pollution. 2020;**267**:115659. DOI: 10.1016/j.envpol.2020.115659

[30] Pazos RS, Suarez JC, Gomez N. Study of the plastisphere: Biofilm development and presence of fecal indicator bacteria on microplastics from the Río de la Plata estuary. Ecosistems. 2020;**29**(3):2069. DOI: 10.7818/ECOS.2069

[31] Galloway TS. Micro- and Nano-Plastics and Human Health. Cham, Switzerland: Springer; 2015. pp. 343-366. DOI: 10.1007/978-3-319-16510-3_13

[32] Todd G, Alan B, Flemming C, Albert K, Shirley P, Sandra W, et al. Dietary and inhalation exposure to nano- and microplastic particles and potential implications for human health. World Health Organization. 2022. DOI: 10.13140/RG.2.2.27459.07200

[33] Hartmann NB, Rist S, Bodin J, Jensen LH, Schmidt SN, Mayer P, et al. Microplastics as vectors for environmental contaminants: Exploring sorption, desorption, and transfer to biota. Integrated Environmental Assessment and Management. 2017a;**13**(3):488-493. DOI: 10.1002/ieam.1904

[34] Miao L, Wang P, Hou J, Yao Y, Liu Z, Liu S, et al. Distinct community structure and microbial functions of biofilms colonizing microplastics. Science of the Total Environment. 2019;**650**:2395-2402

[35] Napper IE, Bakir A, Rowland SJ, Thompson RC. Characterisation, quantity and sorptive properties of microplastics extracted from cosmetics. Mar Pollut Bull. 2015;**15;99**(1-2):178-185. DOI: 10.1016/j.marpolbul.2015.07.029

[36] Hartmann NB, Rist S, Bodin J, Jensen LHS, Schmidt SN, Mayer P, et al. Microplastics as vectors for environmental contaminants: exploring sorption, desorption, and transfer to biota. Integrated Environment Assessment Management. 2017;**13**:488-493. DOI: 10.1002/ieam.1904

[37] Du Y, Liu X, Dong X, Yin Z. A review on marine plastisphere: Biodiversity, formation, and role in degradation. Computational and structural. Biotechnology Journal. 2022;**20**:975-988. DOI: 10.1016/j.csbj.2022.02.008

[38] Lu L, Wan Z, Luo T, Fu Z, Jin Y. Polystyrene microplastics induce gut microbiota dysbiosis and hepatic lipid metabolism disorder in mice. The Science of the Total Environment. 2018a;**631-632**:449-458. DOI: 10.1016/j.scitotenv.2018.03.051

[39] Wang J, Tan Z, Peng J, Qiu Q, Li M. The behaviors of microplastics in the marine environment. Marine Environmental Research. 2016;**113**(7-17):0141-1136. DOI: 10.1016/marenvres.2015.10.014

[40] Wu B, Wu X, Liu S, Wang Z, Chen L. Dependent effects of polystyrene microplastics on cytotoxicity and efflux pump inhibition in human Caco-2 cells. Chemosphere. 2019;**221**:333-341. DOI: 10.1016/j.chemosphere.2019.01.056

[41] Cao Y, Zhao M, Ma X, Song Y, Zuo S, Li H, et al. A critical review on the interactions of microplastics with heavy metals: Mechanism and their combined effect on organisms and humans. The Science of the Total Environment. 2021;**788**:147620. DOI: 10.1016/j.scitotenv.2021.147620

[42] Rodrigues JP, Duarte AC, Santos-Echeandía J, Rocha-Santos T. Significance of interactions between microplastics and POPs in the marine environment: a critical overview. Trends in Analytical Chemistry. 2019;**111**:252-260. DOI: 10.1016/j.trac.2018.11.038

[43] Bakir A, O'Connor IA, Rowland SJ, Hendriks AJ, Thompson RC. Relative importance of microplastics as a pathway for the transfer of hydrophobic organic chemicals to marine life. Environmental Pollution. 2016;**219**:56-65. DOI: 10.1016/j.envpol.2016.09.046

[44] de Sá LC, Oliveira M, Ribeiro F, Rocha TL, Futter MN. Studies of the effects of microplastics on aquatic organisms: What do we know and where should we focus our efforts in the future? The Science of the Total Environment. 2018;**645**:1029-1039. DOI: 10.1016/j.scitotenv.2018.07.207

[45] Hüffer T, Weniger AK, Hofmann T. Sorption of organic compounds by aged polystyrene microplastic particles. Environmental Pollution. 2018;**236**:218-225. DOI: 10.1016/j.envpol.2018.01.022

[46] Lin WH, Wang HY, Kuo J, Lo SL. Adsorption and desorption characteristics of heavy metals onto conventional and biodegradable plastics. Chemosphere. 2023;**333**:138920. DOI: 10.1016/j.chemosphere.2023.138920

[47] Joo SH, Liang Y, Kim M, Byun J, Choi H. Microplastics with adsorbed contaminants: Mechanisms and treatment. Environmental Challenges. 2021;**3**:100042. DOI: 10.1016/j.envc.2021.100042

[48] Verla AW, Enyoh CE, Verla EN, Nwarnorh KO. Microplastic-toxic chemical interaction: A review study on quantified levels, mechanism and implication. SN Applied Sciences. 2019;**1**:1400. DOI: 10.1007/s42452-019-1352-0

[49] Gagliano E, Sgroi M, Falciglia PP, FGA V, Roccaro P. Removal of poly- and perfluoroalkyl substances (PFAS) from water by adsorption: Role of PFAS chain length, effect of organic matter and challenges in adsorbent regeneration. Water Research. 2020b;**171**:115381. DOI: 10.1016/j.watres.2019.115381

[50] Huang Y, Qing X, Wang W, Han G, Wang JY. Mini-review on current studies of microplastics in the air: Analytical methods, occurrence, sources, fate and potential risk to humans. TrAC Trends in Analytical Chemistry. 2020;**125**:115821. DOI: 10.1016/j.trac.2020.115821

[51] Ateia M, Zheng T, Calace S, Tharayil N, Pilla S, Karanfil T. Sorption behavior of real microplastics (MPs): Insights for organic micropollutants adsorption on a large set of well-characterized MPs. Science of the Total Environment. 2020;**720**:137634. DOI: 10.1016/j.scitotenv.2020.137634

[52] Zettler ER, Mincer TJ, Amaral-Zettler LA. Life in the "plastisphere": Microbial communities on plastic marine debris. Environmental Science & Technology. 2013;**47**:7137-7146. DOI: 10.1021/es401288x

[53] Liu S, Shi J, Wang J, Dai Y, Li H, Li J, et al. Interactions between microplastics and heavy metals in aquatic environments: A review. Frontiers in Microbiology. 2021;**12**:652520. DOI: 10.3389/fmicb.2021.652520

[54] Carpenter EJ, Smith KL Jr. Plastics on the Sargasso Sea surface. Science. 1972;**175**(4027):1240-1241. DOI: 10.1126/science.175.4027.1240

[55] Amaral-Zettler LA, Zettler ER, Mincer TJ. Ecology of the plastisphere. Nature Reviews Microbiology. 2020;**18**(3):139-151

[56] Frère L, Maignien L, Chalopin M, Huvet A, Rinnert E, Morrison H, et al. Microplastic bacterial communities in the bay of Brest: Influence of polymer type and size. Environmental Pollution. 2018;**242**:614-625. DOI: 10.1016/j.envpol.2018.07.023

[57] Wilkes RA, Aristilde L. Degradation and metabolism of synthetic plastics and associated products by pseudomonas sp.: Capabilities and challenges. Journal of Applied Microbiology. 2017;**123**:582-593. DOI: 10.1111/jam.13472

[58] Kirstein IV, Kirmizi S, Wichels A, Garin-Fernandez A, Erler R, Löder M, et al. Dangerous hitchhikers? Evidence for potentially pathogenic vibrio spp. on microplastic particles. Marine Environmental Research. 2016;**120**:1-8. DOI: 10.1016/j.marenvres.2016.07.004

[59] Miao L, Wang C, Hou J, Yao Y, Liu Z, Liu S, et al. Distinct microbial metabolic activities of biofilms colonizing microplastics in three freshwater ecosystems. Journal of Hazardous Materials. 2021;**403**:123577

[60] Zhang F, Wang Z, Vijver MG, Peijnenburg WJ. Theoretical investigation on the interactions of microplastics with a SARS-CoV-2 RNA fragment and their potential impacts on viral transport and exposure. Science of the Total Environment. 2022;**842**:156812. DOI: 10.1016/j.scitotenv.2022.156812

[61] Tanaka K, Takada H, Yamashita R, Mizukawa K, Fukuwaka M, Watanuki Y. Accumulation of plastic-derived chemicals in tissues of seabirds ingesting marine plastics. Marine Pollution Bulletin. 2013;**69**:219-222. DOI: 10.1016/j.marpolbul.2012.12.010

[62] Atugoda T, Vithanage M, Wijesekara H, Bolan N, Sarmah AK, Bank MS, et al. Interactions between microplastics, pharmaceuticals and personal care products: Implications for vector transport. Environment International. 2021;**149**:106367. DOI: 10.1016/j.envint.2020.106367

[63] Brennecke D, Duarte B, Paiva F, Caçador I, Canning-Clode J. Microplastics as vector for heavy metal contamination from the marine environment. Estuarine, Coastal and Shelf Science. 2016;**178**:189-195. DOI: 10.1016/j.ecss.2015.12.003

[64] Guan J, Qi K, Wang J, Wang W, Wang Z, Lu N, et al. Microplastics as an emerging anthropogenic vector of trace metals in freshwater: Significance of biofilms and comparison with natural substrates. Water Research. 2020;**184**:116205. DOI: 10.1016/j.watres.2020.116205

[65] Yu Y, Chen H, Hua X, Dang Y, Han Y, Yu Z, et al. Polystyrene microplastics (PS-MPs) toxicity induced oxidative stress and intestinal injury in nematode Caenorhabditis elegans. Science of the Total Environment. 2020;**726**:138679. DOI: 10.1016/j.scitotenv.2020.138679

[66] Gao Y, Yang F, Liu J, Xie W, Zhang L, Chen Z, et al. Genome-wide

identification of metal tolerance protein genes in populus trichocarpa and their roles in response to various heavy metal stresses. International Journal of Molecular Science. 2020;**29;21**(5):1680. DOI: 10.3390/ijms21051680

[67] Boiteux V, Dauchy X, Bach C, Colin A, Hemard J, Sagres V, et al. Concentrations and patterns of perfluoroalkyl and polyfluoroalkyl substances in a river and three drinking water treatment plants near and far from a major production source. The Science of the Total Environment. 2017;**583**:393-400. DOI: 10.1016/j.scitotenv.2017.01.079

[68] Koelmans AA, Mohamed Nor NH, Hermsen E, Kooi M, Mintenig SM, De France J. Microplastics in freshwaters and drinking water: Critical review and assessment of data quality. Water Research. 2019;**155**:410-422. DOI: 10.1016/j.watres.2019.02.054

[69] Pérez-Guevara F, Roy PD, Elizalde-Martínez I, Kutralam-Muniasamy G, Shruti VC. Human exposure to microplastics from urban decentralized pay-to-fetch drinking-water refill kiosks. The Science of the Total Environment. 2022;**848**:157722. DOI: 10.1016/j.scitotenv.2022.157722

[70] Sánchez-Hernández LJ, Ramírez-RomeroP,Rodríguez-GonzálezF, Ramos-Sánchez VH, Márquez Montes RA, Romero-Paredes Rubio H, et al. Seasonal evidences of microplastics in environmental matrices of a tourist dominated urban estuary in Gulf of Mexico, Mexico. Chemosphere. 2021;**277**:130261. DOI: 10.1016/j.chemosphere.2021.130261

[71] Xu S, Ma J, Ji R, Pan K, Miao AJ. Microplastics in aquatic environments: Occurrence, accumulation, and biological effects. The Science of the Total Environment. 2020;**703**:134699. DOI: 10.1016/j.scitotenv.2019.134699

[72] Fenton SE, Ducatman A, Boobis A, DeWitt JC, Lau C, Ng C, et al. Per- and Polyfluoroalkyl substance toxicity and human health review: Current state of knowledge and strategies for informing future research. Environmental Toxicology and Chemistry. 2021;**40**(3):606-630. DOI: 10.1002/etc.4890

[73] Zarus GM, Muianga C, Hunter CM, Pappas RS. A review of data for quantifying human exposures to micro and nanoplastics and potential health risks. The Science of the Total Environment. 2021;**756**:144010. DOI: 10.1016/j.scitotenv.2020.144010

[74] Sun H, Chen N, Yang X, Xia Y, Wu D. Effects induced by polyethylene microplastics oral exposure on colon mucin release, inflammation, gut microflora composition and metabolism in mice. Ecotoxicology and Environmental Safety. 2021a;**220**:112340. DOI: 10.1016/j.ecoenv.2021.112340

[75] Mei LY, Zhou Y, Tsang DCW, Beiyuan J, Song L, She J, et al. Emergent thallium exposure from uranium mill tailings. Journal of Hazardous Materials. 2021. ISSN 0304-3894;**407**. DOI: 10.1016/j.jhazmat.2020.124402

[76] Shyong EQ, Lu Y, GoldsteinA,LebwohlM,WeiH.Synergistic enhancement of H2O2 production in human epidermoid carcinoma cells by benzo[a]pyrene and ultraviolet a radiation. Toxicology and Applied Pharmacology. 2003;**188**(2):104-109. DOI: 10.1016/s0041-008x (03)00018-8

[77] Xia Q, Chiang H-M, Yin J-J. UVA photoirradiation of benzo[a]pyrene metabolites: Induction of cytotoxicity, reactive oxygen species, and lipid

peroxidation. Toxicology and Industrial Health. 2015;**31**:898-910. DOI: 10.1177/0748233713484648

[78] Yadav IC, Devi NL, Syed JH, Cheng Z, Li J, Zhang G, et al. Current status of persistent organic pesticides residues in air, water, and soil, and their possible effect on neighboring countries: A comprehensive review of India. The Science of the Total Environment. 2015;**511**:123-137. DOI: 10.1016/j.scitotenv.2014.12.041

[79] Mo Q, Yang X, Wang J, Xu H, Li W, Fan Q, et al. Adsorption mechanism of two pesticides on polyethylene and polypropylene microplastics: DFT calculations and particle size effects. Environmental Pollution. 2021;**291**:118120. DOI: 10.1016/j.envpol.2021.118120

[80] Lan T, Wang T, Cao F, Yu C, Chu Q, Wang F. A comparative study on the adsorption behavior of pesticides by pristine and aged microplastics from agricultural polyethylene soil films. Ecotoxicology and Environmental Safety. 2021;**209**:111781. DOI: 10.1016/j.ecoenv.2020.111781

[81] Junaid M, Wang J. Interaction of nanoplastics with extracellular polymeric substances (EPS) in the aquatic environment: A special reference to eco-corona formation and associated impacts. Water Research. 2021;**201**:117319. DOI: 10.1016/j.watres.2021.117319

[82] Wang Y, Yang Y, Liu X, Zhao J, Liu R, Xing B. Interaction of microplastics with antibiotics in aquatic environment: Distribution, adsorption, and toxicity. Environmental Science & Technology. 2021;**55**(23):15579-15595. DOI: 10.1021/acs.est.1c04509

[83] Li J, Zhang K, Zhang H. Adsorption of antibiotics on microplastics. Environmental Pollution. 2018;**237**:460-467. DOI: 10.1016/j.envpol.2018.02.050

[84] Rochman CM. Microplastics research from sink to source. Science. 2018;**360**:28-29. DOI: 10.1126/science.aar7734

[85] Waring H, Robert R, Harris M, Mitchell SC. Plastic contamination of the food chain: A threat to human health? Maturitas. 2018;**115**:64-68. DOI: 10.1016/j.maturitas.2018.06.010

[86] Llorca M, Farré M. Current insights into potential effects of micro-Nanoplastics on human health by in-vitro tests. Frontiers in Toxicology. 2021;**3**:752140. DOI: 10.3389/ftox.2021.752140

[87] Alimi OS, Budarz JF, Hernández LM, Tufenkji N. Microplastics and nanoplastics in aquatic environments: Aggregation, deposition, and enhanced contaminant transport. Environmental Science & Technology. 2018;**52**(4):1704-1724. DOI: 10.1021/acs.est.7b05559

[88] [CDC] Centers for Disease Control and Prevention. 2019.

[89] Cai L, Wang J, Peng J, Tan Z, Zhan Z, Tan X, et al. Characteristics of microplastics in the atmospheric fallout from Dongguan city, China: Preliminary research and first evidence. Environmental Science and Pollution Research International. 2017;**24**(32):24928-24935. DOI: 10.1007/s11356-019-06979-x

[90] Waite HR, Donnelly MJ, Walters LJ. Quantity and types of microplastics in the organic tissues of the eastern oyster Crassostrea virginica and Atlantic mud crab Panopeus herbstii from a Florida estuary. Marine Pollution Bulletin. 2018;**129**(1):179-185. DOI: 10.1016/j.marpolbul.2018.02.026

[91] Frias J, Nash R. Microplastics: Finding a consensus on the definition. Marine Pollution Bulletin. 2019;**138**:145-147. DOI: 10.1016/j.marpolbul.2018.11.022

[92] Cox DK, Covernton AG, Davies HL, Dower JF, Juanes F, Dudas SD. Consumption of microplastics. Environmental Science & Technology. 2019;**53**:7068-7074. DOI: 10.1021/acs.est.9b01517

[93] Oßmann BE. Microplastics in drinking water? Present state of knowledge and open questions. Current Opinion in Food Science. 2021;**41**:44-51. DOI: 10.1016/j.cofs.2021.02.011

[94] Yee S-LM, Ling-Wei H, Chin-King L, Wei-Meng L, Shew-Fung W, Kok YY, et al. Impact of microplastics and nanoplastics on human health. Nanomaterials. 2021;**11**:496. DOI: 10.3390/nano11020496

[95] Hernández LM, Xu EG, Larsson HC, Tahara R, Maisuria UV, Tufenkji N. Plastic teabags release billions of microparticles and nanoparticles into tea. Environmental Science & Technology. 2019;**53**:12300-12310. DOI: 10.1021/acs.est.9b02540

[96] Mason SA, Victoria G, Joseph WY, Neratko. Synthetic polymer contamination in bottled water. Frontiers in Chemistry. 2018;**6**:407. DOI: 10.3389/fchem.2018.00407

[97] Santillo D, Molinero K, Johnston P. Microplastics as contaminants in commercially important seafood species. Integrated Environmental Assessment and Management. 2017;**13**(3):516-521. DOI: 10.1002/ieam.1909

[98] Du F, Cai H, Zhang Q, Chen Q, Shi H. Microplastics in take-out food containers. Journal of Hazard Materials. 2020;**15**(399):122969. DOI: 10.1016/j.jhazmat.2020.122969

[99] Deng Y, Zhang Y, Lemos B, Ren H. Tissue accumulation of microplastics in mice and biomarker responses suggest widespread health risks of exposure. Scientific Reports. 2017;**7**:46687. DOI: 10.1038/srep46687

[100] Revel M, Châtel A, Mouneyrac C. Micro(nano)plastics: A threat to human health? Current Opinion in Environmental Science & Health. 2018;**1**:17-23. DOI: 10.1016/j.coesh.2017.10.003

[101] Campanale C, Massarelli C, Savino I, Locaputo V, Uricchio VF. A detailed review study on the potential effects of microplastics and additives of concern on human health. International Journal of Environmental Research and Public Health. 2020;**17**(4):1212. DOI: 10.3390/ijerph17041212

[102] Rist S, Almroth BC, Hartmann NB, Karlsson TM. Perspective on early communications concerning human health aspects of microplastics. The Science of the Total Environment. 2018;**626**:720-726. DOI: 10.1016/j.scitotenv.2018.01.092

[103] Van Raamsdonk L, Van Der Zande M, Koelmans AA, Hoogenboom R, Peters RJB, Groot MJ, et al. Current insights into monitoring, bioaccumulation, and potential health effects of microplastics present in the food chain. Food. 2020;**9**(1):72. DOI: 10.3390/foods9010072

[104] Auta HS, Emenike CU, Fauziah SH. Distribution and importance of microplastics in the marine environment: a review of the sources, fate, effects, and potential solutions. Environment International. 2017;**102**:165-176

[105] Correia-Prata J. Airborne microplastics: Consequences to human health? Environmental Pollution. 2018;**234**:115-126. DOI: 10.1016/J.SOILBIO.2017.10.027

[106] Wright SL, Kelly FJ. Plastic and human health: A micro issue? Environmental Science & Technology. 2017;**51**(12):6634-6647. DOI: 10.1021/acs.est.7b00423

[107] Amato-Lourenço LF, Carvalho-Oliveira R, Júnior GR, Dos Santos GL, Ando RA, Mauad T. Presence of airborne microplastics in human lung tissue. Journal of Hazardous Materials. 2021;**15**(416):126124. DOI: 10.1016/j.jhazmat.2021.126124

[108] Dris R, Gasperi J, Mirande C, Mandín C, Guerrouache M, Langlois V, et al. First overview of textile fibers, including microplastics, in indoor and outdoor environments. Environmental Pollution. 2017;**221**:453-458. DOI: 10.1016/j.envpol.2016.12.013

[109] Hahladakis JN, Velis CA, Weber R, Iacovidou E, Purnell P. An overview of chemical additives present in plastics: Migration, release, fate and environmental impact during their use, disposal and recycling. Journal of Hazardous Materials. 2018;**344**:179-199. DOI: 10.1016/j.jhazmat.2017.10.014

[110] Fournier SB, D'errico JN, Adler DS, Kolontzi S, Goedken MJ, Fabris L, et al. Nanopolystyrene translocation and fetal deposition after acute lung exposure during late-stage pregnancy. Particle and Fibre Toxicology. 2020;**17**:55. DOI: 10.1186/s12989-020-00385-9

[111] Da Silva Brito WA, Mutter F, Wende K, Lourenco Cecchini A, Bekeschus ASS. Consequences of nano and microplastic exposure in rodent models: The known and unknown. Particle and Fibre Toxicology. 2022;**19**(1):28. DOI: 10.1186/s12989-022-00473-y

[112] Ragusa A, Svelato A, Santacroce C, Catalano P, Notarstefano V, Carnevali O, et al. Plasticenta: First evidence of microplastics in human placenta. Environment International. 2021;**146**:106274. DOI: 10.1016/j.envint.2020.106274

[113] Schirinzi GF, Pérez-Pomeda I, Sanchis J, Rossini C, Farre M, Barceló D. Cytotoxic effects of commonly used nanomaterials and microplastics on cerebral and epithelial human cells. Environmental Research. 2017;**159**:579-587. DOI: 10.1016/j.envres.2017.08.043

[114] Allen S, Allen D, Phoenix VR, Le Roux G, Jimenez PD, Simonneau A, et al. Atmospheric transport and deposition of microplastics in a remote mountain catchment. Nature Geoscience. 2019;**12**:339-344. DOI: 10.1038/s41561-019-0335-5

[115] Enyoh CE, Shafea L, Wirnkor-Verla A, Ngozi-Verla E, Qingyue W, Chowdhury T, et al. Microplastics exposure routes and toxicity studies to ecosystems: An overview. Environmental Analysis Health and Toxicology. 2020;**35**:e2020004. DOI: 10.5620/eaht.e2020004

[116] Schneider M, Stracke F, Hansen S, Schaefer UF. Nanoparticles and their interactions with the dermal barrier. Dermatoendocrinology. 2009;**1**:197-206. DOI: 10.4161/derm.1.4.9501

[117] Vogt A, Rancan F, Ahlberg S, Nazemi B, Sik Choe C, Darvin ME, et al. Interaction of dermatologically relevant nanoparticles with skin cells and skin. Beilstein Journal of Nanotechnology. 2014;**5**:2363-2373. DOI: 10.3762/bjnano.5.245

[118] Lai H, Liu X, Qu M. Nanoplastics and Human Health: Hazard Identification

and Biointerface. Nanomaterials (Basel). 2022;**11;12**(8):1298. DOI: 10.3390/nano12081298

[119] Lademann J, Richter H, Schanzer S, Knorr F, Meinke M, Sterry W, et al. Penetration and storage of particles in human skin: perspectives and safety aspects. European Journal of Pharmaceutics and Biopharmaceutics. 2011;**77**(3):465-468. DOI: 10.1016/j.ejpb.2010.10.015

[120] Pironti C, Ricciardi M, Motta O, Miele Y, Proto A, Montano L. Microplastics in the Environment: Intake through the Food Web, Human Exposure and Toxicological Effects. Toxics. 2021;**9**(9):224. DOI: 10.3390/toxics9090224

[121] Liao YL, Yang JY. Microplastic serves as a potential vector for Cr in an in-vitro human digestive model. Science of the Total Environment. 2020;**10**(703):134805. DOI: 10.1016/j.scitotenv.2019.134805

[122] Stock V, Bohmert L, Lisicki E, Block R, Cara-Carmona J, Kim-Pack L, et al. Uptake and effects of orally ingested polystyrene microplastic particles in vitro and in vivo. Archives of Toxicology. 2019;**93**:1817-1833. DOI: 10.1007/s00204-019-02478-7

[123] Fournier E, Etienne-Mesmin L, Grootaert C, Jelsbak L, Syberg K, Blanquet-Diot S, et al. Microplastics in the human digestive environment: A focus on the potential and challenges facing in vitro gut model development. Journal of Hazardous Materials. 2021;**415**:1-15. DOI: 10.1016/j.jhazmat.2021.125632

[124] Ohno H. Intestinal M cells. Journal of Biochemistry. 2016;**159**(2):151-160. DOI: 10.1093/jb/mvv121

[125] Lunov O, Syrovets T, Loos C, Beil J, Delacher M, Tron K, et al. Differential uptake of functionalized polystyrene nanoparticles by human macrophages and a Monocytic cell line. ACS Nano. 2011;**5**(3):1657-1669. DOI: 10.1021/nn2000756

[126] Ding Y, Zhang R, Li B, Du Y, Li J, Tong X, et al. Distribution of polystyrene nanoplastics in mice and their entry, transport, and cytotoxicity to ges-1 cells. Environmental Pollution. 2021;**280**:116974. DOI: 10.1016/j.envpol.2021.116974

[127] Li S, Shi M, Wang Y, Xiao Y, Cai D, Xiao F. Keap1-Nrf2 pathway up-regulation via hydrogen sulfide mitigates polystyrene microplastics induced-hepatotoxic effects. Journal of Hazardous Materials. 2021;**15**(402):123933. DOI: 10.1016/j.jhazmat.2020.123933

[128] Ju H, Zhu D, Qiao M. Effects of polyethylene microplastics on the gut microbial community, reproduction and avoidance behaviors of the soil springtail, *Folsomia candida*. Environmental Pollution. 2018;**247**:890-897. DOI: 10.1016/j.envpol.2019.01.097

[129] Liu Z, Yu P, Cai M, Wu D, Zhang M, Chen M, et al. Effects of microplastics on the innate immunity and intestinal microflora of juvenile *Eriocheir sinensis*. The Science of the Total Environment. 2019;**685**:836-846. DOI: 10.1016/j.scitotenv.2019.06.265

[130] Lu K, Lai KP, Stoeger T, Ji S, Lin Z, Lin X, et al. Detrimental effects of microplastic exposure on normal and asthmatic pulmonary physiology. Journal of Hazardous Materials. 2021;**15**(416):126069. DOI: 10.1016/j.jhazmat.2021.126069

[131] Jin Y, Lu L, Tu W, Luo T, Fu Z. Impacts of polystyrene microplastic

on the gut barrier, microbiota and metabolism of mice. Science Total Environment. 2019;**1**(649):308-317. DOI: 10.1016/j.scitotenv.2018.08.353

[132] Luo T, Wang C, Pan Z, Jin C, Fu Z, Jin Y. Maternal polystyrene microplastic exposure during gestation and loctation altered metabolic homeostasis in the dams and their F1 and F2 offspring. Environmental Science & Technology. 2019;**53**:10978-10992. DOI: 10.1021/acs.est.9b03191

[133] Banerjee A, Shelver WL. Micro- and nanoplastic induced cellular toxicity in mammals. A review. Science of the Total Environment. 2021;**755**:142518. DOI: 10.1016/j.scitotenv.2020.142518

[134] Sadler D, Brunner FS, Plaistow SJ. Temperature and clone-dependent effects of microplastics on immunity and life history in Daphnia magna. Environmental Pollution. 2019;**255**(Pt 1):113178. DOI: 10.1016/j.envpol.2019.113178

[135] Brandts I, Teles M, Gonçalves AP, Barreto A, Franco-Martinez L, Tvarijonaviciute A, et al. Effects of nanoplastics on mytilus galloprovincialis after individual and combined exposure with carbamazepine. The Science of the Total Environment. 2018;**643**:775-784. DOI: 10.1016/j.scitotenv.2018.06.257

[136] Revel M, Lagarde F, Perrein-Ettajani H, Bruneau M, Akcha F, Sussarellu R, et al. Tissue-specific biomarker responses in the blue mussel *Mytilus* spp. exposed to a mixture of microplastics at environmentally relevant concentrations. Frontiers in Environmental Science. 2019;**7**. DOI: 10.3389/fenvs.2019.00033/full

[137] Greven AC, Merk T, Karagoz F, Mohr K, Clapper M, Jovanovic B, et al. Polycarbonate and polystyrene nanoplastic particles act as stressors to the innate immune system of fathead minnow (Pimephales promelas). Environmental Toxicology and Chemistry. 2016;**35**(12):3093-3100. DOI: 10.1002/etc.3501

[138] Park EJ, Han JS, Park EJ, Seong E, Lee GH, Kim DW, et al. Repeated-oral dose toxicity of polyethylene microplastics and the possible implications on reproduction and development of the next generation. Toxicology Letters. 2020;**324**:75-85. DOI: 10.1016/j.toxlet.2020.01.008

[139] Hirt N, Body-Malapel M. Immunotoxicity and intestinal effects of nano- and microplastics: A review of the literature. Particle and Fibre Toxicology. 2020;**17**(1):57. DOI: 10.1186/s12989-020-00387-7

[140] Mu Q, Kirby J, Reilly CM, Luo XM. Leaky gut as a danger signal for autoimmune diseases. Frontiers in Immunology. 2017;**8**:598. DOI: 10.3389/fimmu.2017.00598

[141] Schwabl P, Köppel S, Königshofer P, Bucsics T, Trauner M, Reiberger T, et al. Detection of various microplastics in human stool: A prospective case series. Annals of Internal Medicine. 2019;**171**(7):453-457. DOI: 10.7326/M19-0618

[142] Ibrahim YS, Tuan Anuar S, Azmi AA, Wan Mohd Khalik WMA, Lehata S, Hamzah SR, et al. Detection of microplastics in human colectomy specimens. JGH Open. 2020;**5**(1):116-121. DOI: 10.1002/jgh3.12457

[143] Silva CJM, Beleza S, Campos D, Soares AMVM, Patrício Silva AL, Pestana JLT, et al. Immune response triggered by the ingestion of polyethylene microplastics in the dipteran larvae Chironomus riparius. Journal of Hazardous Materials.

2021;**15**(414):125401. DOI: 10.1016/j.jhazmat.2021.125401

[144] Sommer F, Nookaew I, Sommer N, Fogelstrand P, Backhed F. Site-specific programming of the host epithelial transcriptome by the gut microbiota. Genome Biology. 2015;**16**:62. DOI: 10.1186/s13059-015-0614-4

[145] Spadoni I, Zagato E, Bertocchi A, Paolinelli R, Hot E, Di Sabatino A, et al. A gut vascular barrier controls the systemic dissemination of bacteria. Science. 2015;**350**(6262):830-834. DOI: 10.1126/science.aad0135

[146] Mouries J, Brescia P, Silvestri A, Spadoni I, Sorribas M, Wiest R, et al. Microbiota-driven gut vascular barrier disruption is a prerequisite for non-alcoholic steatohepatitis development. Journal of Hepatology. 2019;**71**(6):1216-1228. DOI: 10.1016/j.jhep.2019.08.005

[147] Cheng C, Tan J, Qian W, Zhang L, Hou X. Gut inflammation exacerbates hepatic injury in the high-fat diet-induced NAFLD mouse: Attention to the gut-vascular barrier dysfunction. Life Sciences. 2018;**15**(209):157-166.09. DOI: 10.1016/j.lfs.2018.08.017

[148] Zhang Y, Kang S, Allen S, Allen D, Gao T, Sillanpää M. Atmospheric microplastics: A review on current status and perspectives. Earth-Science Reviews. 2020;**203**:103118. DOI: 10.1016/j.earscirev.2020.103118

[149] Lu K, Zhan D, Fang Y, Li L, Chen G, Chen S, et al. Microplastics, potential threat to patients with lung diseases. Frontier Toxicology. 2022;**28**(4):958414. DOI: 10.3389/ftox.2022.958414. Erratum in: Frontier Toxicology. 2022;**20**(4):1119994

[150] Fan Z, Xiao T, Luo H, Chen D, Lu K, Shi W, et al. A study on the roles of long non-coding RNA and circular RNA in the pulmonary injuries induced by polystyrene microplastics. Environment International. 2022;**163**:107223. DOI: 10.1016/j.envint.2022.107223

[151] Lim X. Microplastics are everywhere - but are they harmful? Nature. 2021;**593**(7857):22-25. DOI: 10.1038/d41586-021-01143-3

[152] Silva AH, Locatelli C, Filippin-Monteiro FB, Martin F, Liptrott NJ, Zanetti-Ramos BG, et al. Toxicity and inflammatory response in Swiss albino mice after intraperitoneal and oral administration of polyurethane nanoparticles. Toxicology Letters. 2016;**246**:17-27. DOI: 10.1016/j.toxlet.2016.01.018

[153] Zheng H, Wang J, Wei X, Chang L, Liu S. Proinflammatory properties and lipid disturbance of polystyrene microplastics in the livers of mice with acute colitis. The Science of the Total Environment. 2021;**750**:143085. DOI: 10.1016/j.scitotenv.2020.143085

[154] Liss KH, Finck BN. PPARs and nonalcoholic fatty liver disease. Biochimie. 2017;**136**:65-74. DOI: 10.1016/j.biochi.2016.11.009

Chapter 11

Effect of Microplastic on the Human Health

Ahmad K. Jassim

Abstract

Microplastics are defined as a tiny plastic particle with length less than 5 mm that result from commercial product development and the breakdown of larger plastics can be considered as harmful to our environment. The surfaces of plastic tiny fragments can be carrying disease-causing organisms and acts as a vector for diseases. The microplastics can found in water and soil carrying different types of contaminants. The small particles of microplastics serve as carriers for bacteria and persistent organic pollutants, which are considered as biodegradable material that remains active for years. The main definition of microplastics and their source as well as the effect of microplastic waste on the human health with the main solution that helps to eliminate this kind of waste will be explained in this chapter. It was found that the human person eats high quantity of microplastic particles every year through breath air, drinking water, and eating plant crops and animals. On the other hand, sea fish can contain microplastic parts due to drinking and eating sea water with microplastics. Therefore, it is very important to reduce throwing of plastic waste on the ocean and landfill to avoid the effect of microplastic on the human health.

Keywords: microplastic, human health, environment, waste, soil

1. Introduction

Due to increasing the number of populations worldwide, the consumption of plastic was increased too and led to produce and generate microplastic waste. The plastics were entered into all parts of our life from clothing to cleaning products because it is a cheaper, durable, and lightweight material. Large quantities of plastics are released into rivers and oceans, which can be classified into large plastic wastes and small plastic particulates named microplastics [1].

Microplastics are defined as tiny plastic particles that form as a result of the plastic content in consumer products and the breakdown of large plastics. Once they have entered the environment become pollutants, causing harm to the environment and animal health.

Microplastics prove much more challenging to remove due to their small sizes, which are often too small to spot or grab in moving water and can remain active marine contaminants for up to 450 years. The small particles of microplastics serve as carriers for bacteria and persistent organic pollutants, which are toxic organic compounds that take years to degrade. Moreover, they consist of chemical materials that are hazardous to human and animal health in high concentrations. The humans ingest microplastics by eating marine animals that have consumed the material or through drinking water or breathing air. Therefore, it is very important to have idea and information about microplastics and how to avoid or eliminate their effect on our life.

This chapter will explain the main source and effect of microplastic on the human health.

2. Microplastic

Increasing consumption of plastic products in modern society has caused microplastic contamination in nearly all environmental media. Microplastics are tiny plastic particles that result from both commercial product development and the breakdown of larger plastics. As a pollutant, microplastics can be harmful to the environment and animal health. Microplastics are fragments of any type of plastic less than 5 mm in length, according to the U.S. National Oceanic and Atmospheric Administration and the European Chemicals Agency. Moreover, microplastics can be defined as heterogeneously mixed plastics that include plastic fibers, granules, and fragments with diameters less than 5 mm. They are considered to be emerging contaminants of concern [2].

3. Source of microplastic

The main source of microplastics can be defined as tires, synthetic textiles, marine coatings, road markings, personal care products, plastic pellets, and city dust. Moreover, the majority of microplastics come from household activity with a percentage of 77% and 23% from industrials application [1].

4. Effect of microplastic waste on the human health

Based on the latest global estimate of microplastics, there are 93–236 thousand tons of microplastics floating on the ocean surface, which corresponds to 51 trillion particles. 79% of global plastic waste is stacked in landfills, which makes soil a large microplastic sink. For example, 90% of Swiss floodplain soils have microplastics at depths between 0 and 5 cm [1].

Plastic products are usually manufactured by using new resources basically petroleum-based materials. However, they lost their value during their life cycles due to leakage along the entire value chain such as pellet loss, loss during transportation, and storage of plastic waste and littering as well as combined sewage overflow and poorly designed products, which lead to loss into our environment easily and difficult to recover as shown in **Figure 1**. This leads to contamination of the environment, affecting wildlife and human well-being. A small proportion is recycled for remanufacturing with remainder utilized for energy recovery [3].

The sources of microplastics are mainly classified as either primary or secondary microplastics. Primary microplastics are purposefully manufactured for specific applications, which include cosmetic abrasives, drug vectors, and industrial and engineering applications such as air blasting. These microplastics are usually difficult to remove using sewage disposal technologies and once they enter wastewater, they will ultimately accumulate in the environment. Secondary microplastics originate from larger

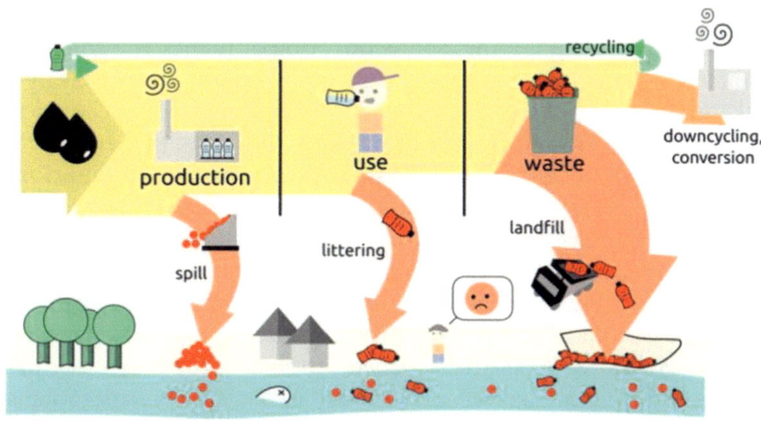

Figure 1.
Life cycle of microplastic and plastic products.

plastics as they are progressively fragmented into smaller pieces by multiple, complex environmental conditions such as wind, waves, temperature, and UV light [2].

The sources of microplastics in the ocean mainly include land-based sources with a percentage of 80% caused by coastal tourism, recreational, commercial fishing, or plastic fishing gear applications and 18% caused by marine vessels and marine industries such as aquaculture and oil rigs. The microplastics enter soil via multiple sources that include landfills, soil amendments, land application of sewage sludge, wastewater irrigation, compost and organic fertilizer, residues of agricultural mulching films, tire wear and tear, and atmospheric deposition. **Figure 2** shows source of microplastic waste in soil [2, 4].

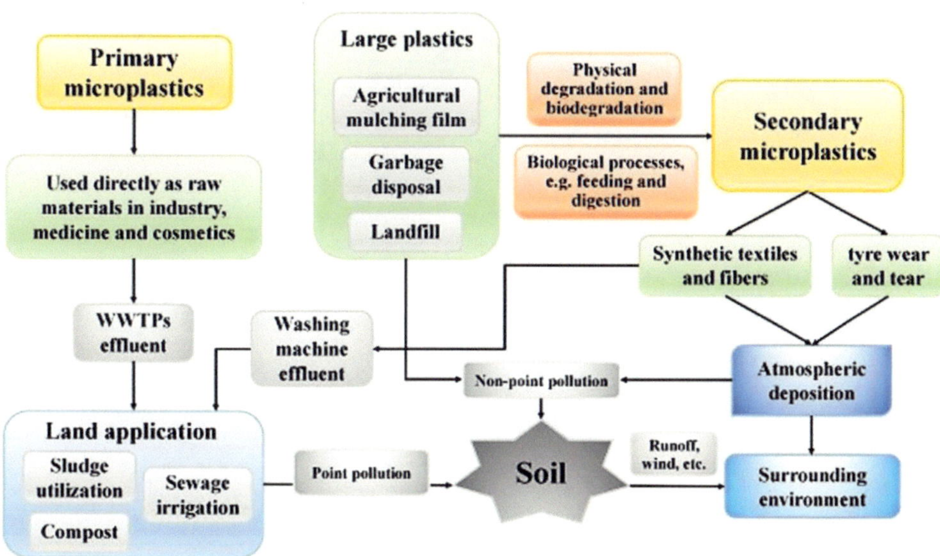

Figure 2.
Source of microplastics in soil [2].

Microplastic waste can be affecting soil by making soil cracking, pores, agronomic practices by plowing and harvesting, root elongation of plants, the activities of ingestion, and egestion of geophagies soil fauna. Moreover, the excavating activities of other soil animals contribute likely to the most vertical transport of microplastics in soil; however, the activities of hunting and life activities of epigenic earthworms, as well as agronomical practices can be facilitating the horizontal distribution of microplastics in soil. Additionally, plastic types can also influence the migration, because those microbeads and microfibers have been proved to show different interactions with soil aggregation, which may exert potentially blocking effects on the transport of microplastics in soil. Furthermore, transport may be influenced by plastic surface properties and eco-corona altered by the process of degradation. Soil nature influences the migration of microplastics, which lies to change the properties of soil such as soil structure and function as well as microbial diversity. This microbial may translate to plant and animal consequences and present potential concerns for food quality and safety, ultimately threatening human health as shown in **Figure 3**.

Salt is mostly produced by the distillation of seawater, which will contained microplastic materials. It is difficult to avoid microplastics in final sea salt products without further purification steps. Thermoplastics such as polyethylene, polypropylene, and polystyrene comprised the majority of microplastics that can be found in the food, which account for more than 50%. Fibers are critical because they are thought to cause toxic effects at lower doses than spherical particles. Fibers including particles classified as "filaments" were dominant in many food items. For example, the fraction of fibers reached almost 100% of microplastics in sea salts and edible tissues of fish and shellfish [5–11].

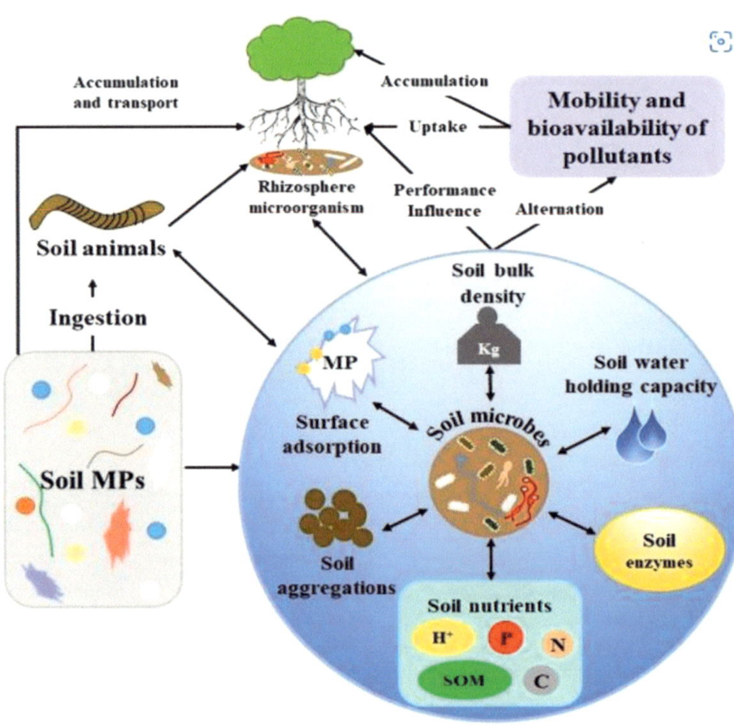

Figure 3.
Impact of microplastics on soil function and properties [2].

Microplastics have been found in different foods such as fish and seafood, table salt, beer, honey and sugar, and tap water. On the other hand, it is found in soil as a result of contamination from items such as discarded packaging or plastic agricultural equipment [12–16].

The studies show that the human person eats at least 50,000 microplastic particles every year because of the infiltrated food chain, drinking water, and breathing air. There are seven types of sources can be considered as food sources of microplastics that includes bottled water, beer, sea food (Shellfish and fish), salt, tee Bag, canned Food, and ready meals. Water in plastic bottles that is used for drinking is one of the worst sources of microplastic, which results in ingesting around 130,000 fragments of microplastic in the human body yearly. The infiltration only gets worse when the bottle is exposed to direct sunlight. The tap water contains tiny plastic bits but the level in bottled water is double than tap water. On the other hand, 1 liter of beer contains 4.05 plastic fibers. Moreover, shellfish is the second major source of microplastic after water because many marine species mistakenly eat plastic debris floating in the ocean. Tiny plastic fibers are present in their entire body including bivalves, which are consumed by humans. In addition, 1 kilogram of sea salt contains 212 particles of microplastic. Furthermore, plastic teabags when dipped in hot water leach microplastics in the cup, therefore, most brands use paper teabags. The biggest health threat with canned foods is the chemical BPA because it is used to harden the plastic, which seeps into the food inside and contaminates it. Finally, the quick meals are usually served in plastic containers, which add a secret ingredient to your diet along with other nutrients microplastic [17].

There is a different source of plastic waste that can be collected from any wildlife in or around rivers, which is exposed to the threats of microplastic pollution. Microplastics can block the gastrointestinal tracts of organisms, or trick them into thinking they do not need to eat, leading to starvation as shown in **Figures 4–6** [18].

Microplastic materials can be carried a range of contaminants such as trace metals and some potentially harmful organic chemicals. The chemical materials can be leaching from the plastic surface into the body, which leads to an increase in the

Figure 4.
Plastic and microplastic waste collected around rivers [18].

Figure 5.
Municipal workers cleaning plastic wastes from river [18].

Figure 6.
Different types of plastic wastes covering the top surface of rivers [18].

potential of toxic effects. Moreover, microplastics can cause cancer because they have carcinogenic properties.

Microplastics were found in tap water, which can be carried on their surfaces disease-causing organisms, and act as a vector for diseases in the environment. It is interacted with soil fauna, affecting their health and soil functions. Moreover, microplastics can be blocked the gastrointestinal tract of small birds and fish as well as they can be caused internally physical damage such as lacerations or irritation to sensitive gastrointestinal tissues.

Researchers show that human exposure to microplastics could lead to oxidative stress, DNA damage, and inflammation, among other health problems. Particularly, when inflammation becomes chronic, this can pave the way to very serious health problems.

The fruit and vegetables have contamination that transpires when the plants suck water that contains microplastic up through their roots. The majority of crops have microplastics such as pears, potatoes, radishes, and lettuce. However, apples and carrots have the highest levels of microplastic particles.

A recent study has found that people eat 5 grams of micro and nano plastics every week. From the most remote depths of the ocean to the deepest section of the lung, microplastics appear to have invaded every bit of our lives, including the human gastrointestinal tract. In laboratory tests, microplastics have been shown to cause damage to human cells, including both allergic reactions, cell death, damaging cells, and inducing inflammatory and immune reactions.

Regarding the microplastic contamination of soil, the ecological and health risk resulting from microplastic exposure was of significant concern. Microplastics may concentrate in human body through various exposure pathways as shown in **Figure** 7 such as inhalation of dust, consumption of food, or direct drinking water contaminated by microplastics [19].

According to European Food Safety Authority (EFSA), still, we do not have enough information and data that explain how microplastics are present in food and how to affect human health and their behavior in our bodies. Therefore, it is very important to make more research to develop new standardization methods to analyze microplastics and determine their effect and risk on the human body. At the moment there is no advice from official authorities indicating that we need to change our eating behavior or avoid certain foods to steer clear of microplastics [13, 14].

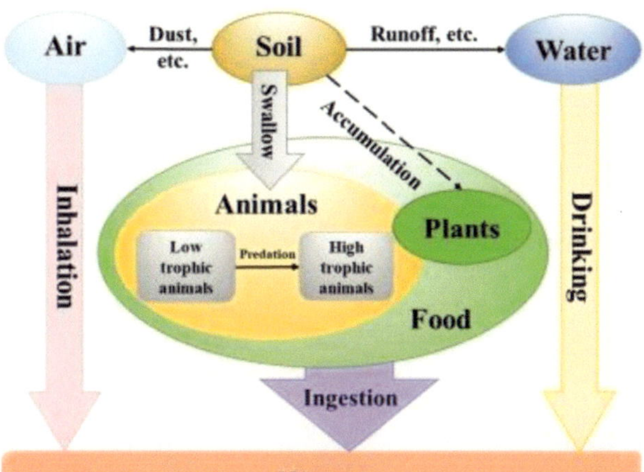

Figure 7.
Various contact pathways of concentration microplastics in human body [19].

5. Solution to eliminate microplastic waste

The ocean cleanup has developed the first scalable solution to efficiently intercept plastic in rivers before it reaches the oceans. By tackling 1000 rivers around the world. To eliminate the microplastic waste effect there are seven things that can be done, which include [20]:

1. Reduce your use of single-use plastics.
2. Support legislation to curb plastic production and waste.
3. Recycle properly.
4. Participate in a beach or river cleanup.
5. Avoid products containing microbeads.
6. Spread the word.
7. Buy a water filter, and stop using bottled water.
8. Buy non-synthetic eco-friendly clothes.
9. Get a laundry ball.
10. Air dry, do not use the dryer.
11. Use public transport, and favor rail infrastructure.
12. Reduce your meat and fish consumption.
13. Use active carbon filter on your tap water
14. Using paper bag instead of plastic bag.

6. Conclusions

Microplastics demonstrate much more challenging to remove due to their small size and huge quantity, which can be remain active for years. The small particles of microplastics serve as carriers for bacteria and persistent organic pollutants, which are considered as biodegradable materials. Moreover, they consist of chemical materials that are hazardous to human and animal health in high concentrations. Many sources of food and landfill have microplastic waste that goes directly or indirectly to the human body by eating food or breathing air, which affect human health. It was found that the human person eats at least 50,000 microplastic particles every year. Still, there is not enough data that help to know how microplastics affect humans. However, there are some solutions to eliminate the effect of microplastics on the humans and animals. Therefore, it is very important to make more researches to develop new standardization methods to analyze microplastics and determine their effect and risk on the human body.

Author details

Ahmad K. Jassim
The State Company for Iron and Steel, Research and Development Department, Basra, Iraq

*Address all correspondence to: ahmadkj1966@yahoo.com

References

[1] Boucher J, Friot D. Primery microplastics in the oceans: A global evaluation of sources. Available from: IUCN+-+Primart+Microplastics+in+the +Oceans-+a+Global+Evalutation+of+So urces,+Julien+Boucher+and+Damien+Fr iot.pdf (squarespace.com)

[2] Jing-JieGuo X-PH, Xiang L, Wang Y-Z, Li Y-W, Li H, Cai Q-Y, et al. Source, migration and toxicology of microplastics in soil. Environment International. 2020;**137**:105263

[3] Eriksen M, Thiel M, Prindiville M, Kiessling T. Microplastic: What Are the Solutions? Fresh water microplastic Book (e Book), the hand book of Environmental Chemistry. 2018;**58**:273-298. Available from: https://link.springer.com/content/pdf/10.1007/978-3-319-61615-5.pdf

[4] Andrady AL. Microplastics in the marine environment. Marine Pollution Bulletin. 2011;**62**:1596-1605. DOI: 10.1016/j.marpolbul.2011.05.030

[5] Qiao R, Deng Y, Zhang S, Wolosker MB, Zhu Q, Ren H, et al. Accumulation of different shapes of microplastics initiates intestinal injury and gut microbiota dysbiosis in the gut of zebrafish. Chemosphere. 2019;**236**:124334. DOI: 10.1016/j.chemosphere.2019.07.065

[6] Ziajahromi S, Kumar A, Neale PA, Leusch FDL. Impact of microplastic beads and fibers on waterflea (Ceriodaphnia dubia) survival, growth, and reproduction: Implications of single and mixture exposures. Environmental Science & Technology. 2017;**51**:13397-13406. DOI: 10.1021/acs.est.7b03574

[7] Au SY, Bruce TF, Bridges WC, Klaine SJ. Responses of Hyalella azteca to acute and chronic microplastic exposures. Environmental Toxicology and Chemistry. 2015;**34**:2564-2572. DOI: 10.1002/etc.3093

[8] Kosuth M, Mason SA, Wattenberg EV. Anthropogenic contamination of tap water, beer, and sea salt. PLoS One. 2018;**13**:e0194970. DOI: 10.1371/journal.pone.0194970

[9] Feng Z, Zhang T, Li Y, He X, Wang R, Xu J, et al. The accumulation of microplastics in fish from an important fish farm and mariculture area, Haizhou Bay, China. Science of The Total Environment. 2019;**696**:133948. DOI: 10.1016/j.scitotenv.2019.133948

[10] Wu F, Wang Y, Leung JYS, Huang W, Zeng J, Tang Y, et al. Accumulation of microplastics in typical commercial aquatic species: A case study at a productive aquaculture site in China. Science of The Total Environment. 2020;**708**:135432. DOI: 10.1016/j.scitotenv.2019.135432

[11] Kwon J-H, Kim J-W, Pham TD, Tarafdar A, Hong S, Chun S-H, et al. Microplastics in food: A review on analytical methods and challenges. International Journal Environment Research Public Health. 2020;**17**(18):6710

[12] Wang J et al. Microplastics as contaminants in the soil environment: A mini-review. Science of the Total Environment. 2019;**691**:848-857

[13] Munoz-Piniero MA. Microplastics: Focus on Food and Health. Publications Office of the European Union; 2018. Available from: https://publications.jrc.ec.europa.eu/repository/handle/JRC110629

[14] European Food Safety Authority, EFSA Panel on Contaminants in the

Food Chain (CONTAM). Presence of microplastics and nano plastics in food, with particular focus on seafood. EFSA Journal. John Wiley and Sons Ltd. 2016;**14**(6):1-30. Available from: https://efsa.onlinelibrary.wiley.com/doi/epdf/10.2903/j.efsa.2016.4501

[15] Ritchie H, Roser M. Plastic Pollution. 2018

[16] The European Parliament. Plastic in the Ocean: The Facts, Effects and New EU Rules. 2021. Available from: https://www.europarl.europa.eu/news/en/headlines/society/20181005STO15110/plastic-in-the-ocean-the-facts-effects-and-new-eu-rules

[17] 7 Foods that Contain Microplastics & How to Avoid them (everythingbagsinc.com). Available from: https://www.everythingbagsinc.com/blog/foods-containing-microplastics

[18] Abu Al-Khasib Municipality. 2022. Available from: https://m.facebook.com/story.php?story_fbid=pfbid02HHSfvFtUzsqE94y3pJR9dLvx8LvsKzB2qzKfy1jBjFKvw7JE5fih8SS6b7KUYauvl&id=100015300409432&sfnsn=mo

[19] Dris R, Gasperi J, Mirande C, Mandin C, Guerrouache M, Langlois V, et al. A first overview of textile fibers, including microplastics, in indoor and outdoor environments. Environmental Pollution. 2017;**221**:453-458. DOI: 10.1016/j.envpol.2016.12.013

[20] Microplastics in food and water – Are they harmful to human health, Microplastics in food and water – are they harmful to human health? Eufic, 2021. Available from: https://www.eufic.org/en/food-safety/article/microplastics-in-food-and-water-are-they-harmful-to-human-health